Science Projects in Electronics

by Edward M. Noll

HOWARD W. SAMS & CO., INC.
THE BOBBS-MERRILL CO., INC.
INDIANAPOLIS · KANSAS CITY · NEW YORK

SECOND EDITION

FIRST PRINTING—1971

Copyright © 1963 and 1971 by Howard W. Sams & Co., Inc., Indianapolis, Indiana 46206. Printed in the United States of America.

All rights reserved. Reproduction or use, without express permission, of editorial or pictorial content, in any manner, is prohibited. No patent liability is assumed with respect to the use of the information contained herein. While every precaution has been taken in the preparation of this book, the publisher assumes no responsibility for errors or omissions. Neither is any liability assumed for damages resulting from the use of the information contained herein.

International Standard Book Number: 0-672-20846-6
Library of Congress Catalog Card Number: 74-157800

Preface

This book has two major objectives: (1) to show basic electronic principles through simple projects and demonstrations, and (2) to help you develop practical construction and testing skills. Simple demonstrations are included in the first chapter to provide a review of basic electric-current fundamentals. Subsequent projects are designed to introduce, in a step-by-step fashion, transistor fundamentals and the concepts of amplification, feedback, oscillation, modulation, detection, and the reception of code signals. Every chapter contains sufficient text material to explain what happens in each of the demonstrations.

An attempt has been made to design all the projects so that inexpensive and readily available parts are used. The projects are planned to be useful as well as educational. For those interested in pursuing the hobby of amateur radio, material has been included on the construction and use of a code-practice oscillator. The final project tells how to build a four-band, transistorized a-m radio receiver.

This book will help you learn many of the fundamentals of electronics in an easy and enjoyable way.

EDWARD M. NOLL

Contents

CHAPTER 1

ELECTRONIC COMPONENTS AND MEASUREMENTS 7

AC and DC Electricity—AC Electricity—The VOM—Equipment and Construction—Using the VOM—Inductors and Transformers—Capacitors and Resonant Circuits

CHAPTER 2

DIODE AND TRANSISTOR 39

Rectifier Operation—Filtering—The Semiconductor Junction—The Bipolar Transistor—DC and AC Operation—The Unipolar or Field-Effect Transistor—FET AC Operation—Rectifier Operation—AC-to-DC Power Supply—Bipolar Transistor Operation

CHAPTER 3

AUDIO AMPLIFIER AND PLAYER 69

Microphone—Phono Cartridge—Audio Amplifier—Speaker Operation—Push-Pull Amplifier—Two-Stage Audio Amplifier—Audio Amplifier at Work

CHAPTER 4

SIMPLE TRANSISTOR RADIO 82

Amplitude Modulation—Receiver Principles—Crystal Radio

CHAPTER 5

AUDIO OSCILLATOR 93

Feedback—Audio Oscillator Types—Tone Oscillator—Code Practice

CHAPTER 6

RADIO-FREQUENCY OSCILLATOR 104

Radio-Frequency Feedback—RF Oscillator Types—Zero-Beating—FET RF Oscillator—Frequency Measurement and Calibration

CHAPTER 7

SIGNAL GENERATOR AND WIRELESS PLAYER 116

Amplitude Modulation—Transistor Modulation Methods—Tone Modulation—Wireless Player

CHAPTER 8

REGENERATIVE RECEIVER AND INTEGRATED-CIRCUIT AUDIO
AMPLIFIER 123

Transistor Detectors—Regenerative Detector—FET Regenerative Detector—Amateur and Short-Wave Radio Reception—Integrated-Circuit Audio Amplifier

INDEX 141

1

Electronic Components and Measurements

In grasping the principles of operation of the various project units, some previous knowledge of electrical fundamentals is helpful, although certainly not mandatory. This chapter reviews important basic electrical principles.

AC AND DC ELECTRICITY

Electric current flow is the motion of tiny negatively charged particles called *electrons*. Electrical energy can be produced in several ways. The battery is an example of a chemical process that produces electron motion.

The amount of current that flows in the external circuit depends on the opposition or *resistance* the path presents to the motion of electrons. The unit of resistance is the *ohm;* the higher the ohmic value of the path, the smaller is the current flow. Thus the amount of current flow is dependent on both the voltage across and the resistance along the path. This relationship can be stated in a simple formula referred to as "Ohm's law":

$$I = \frac{E}{R}$$

where,
    I is the current in amperes,
    E is the emf in volts,
    R is the resistance in ohms.

Table 1-1. Electrical Units and Symbols

Characteristic	Symbol	Unit
Voltage	E	Volt
Current	I	Ampere
Resistance	R	Ohm
Power	P	Watt

The formula states that the current through an electrical circuit is directly proportional to the voltage and inversely proportional to the resistance. Ohm's law can also be stated in the following ways:

$$R = \frac{E}{I}, \text{ or } E = IR$$

Still another important electrical characteristic is *power*. Power represents the rate at which work is done. In this case it represents the rate at which electrical energy is used in the circuit. The higher the current flow and the greater the resistance in the circuit, the more energy is dissipated and, therefore, the greater is the electrical power. The unit of electrical power is the *watt*. Table 1-1 and Table 1-2 summarize the various electrical units and common electrical prefixes.

Table 1-2. Electrical Prefixes and Multipliers

Prefix	Multiplier
Kilo	1000 (One thousand times)
Mega (or Meg)	1,000,000 (One million times)
Milli	1/1000 (One thousandth part of)
Micro	1/1,000,000 (One millionth part of)

AC ELECTRICITY

Electrical energy can be produced by rotating a conductor in a magnetic field. The current flow is first in one direction and then in the other direction. In other words, this type of generator produces alternating current flow, or ac electricity.

As long as the conductor continues to rotate, the current continues to alternate, generating a current wave called a *sine wave*. This is the basic ac waveform. The ac electricity at the outlets in your house varies in essentially this manner. The number of sine waves formed each second by an ac generator is called the *frequency*.

The voltage and current ratings for ac circuits are based on a so-called *effective,* or *root-mean-square* (*rms*) value. A sine wave of so many volts is one that will cause the power dissipation in the resistance of the circuit to be the same as though an equal number of dc volts were applied to the circuit. Likewise the power dissipated in the resistance of the circuit from a given amount of ac current (rms value) would be the same as though the same value of dc current flowed through the same resistance.

Ac voltages and current are sometimes given in terms of peak, average, or peak-to-peak value. These values are indicated in Fig. 1-1.

Ohm's law is basic to ac and dc circuits alike. When a resistance is placed across an ac power source, the relation of voltage and current is the same as for a similar dc circuit. The formulas given previously for voltage, current, and resistance remain true, provided consistent units are used; that is, if rms voltage is used, rms current must be used, etc.

In a circuit containing resistance only, the power is the product of the rms voltage and current. Unless stated otherwise, ac voltage and current are usually given in rms values.

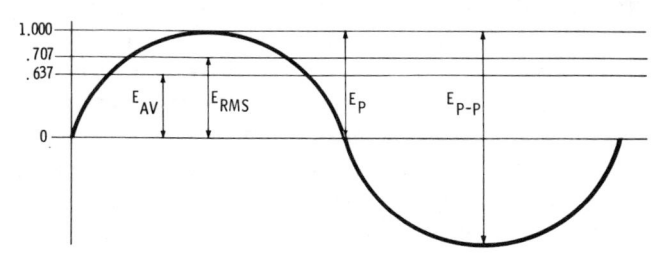

Fig. 1-1. Sine-wave voltage values.

THE VOM

The most common test instrument used for general measurement, testing, and troubleshooting of electronic circuits is the volt-ohm-milliammeter (vom). As its name implies, it can be used to read volts, ohms, and milliamperes. Thus it is used for checking voltage, resistance, and current in electronic circuits. Instruments suitable for general electronic-service use fall in the $10 to $30 price range. Such an instrument is used throughout the projects in this book for making measurements and adjustments and for demonstrating electronic circuit operation.

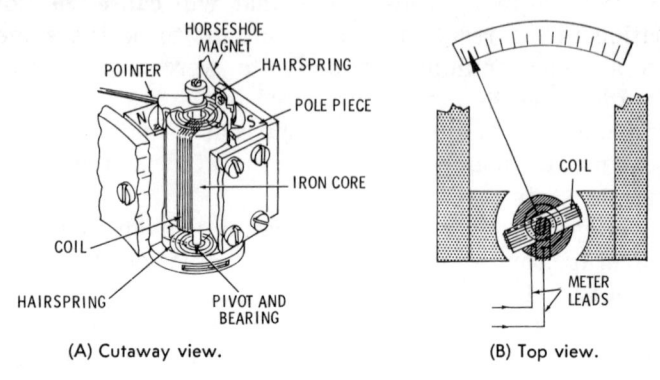

Fig. 1-2. Basic meter movement.

The vom is fundamentally a current-indicating device, although it is used more often to measure voltage and resistance. The meter movement (Fig. 1-2) consists of a permanent magnet and a core-wound coil that is pivot-mounted in the magnetic field. Its principle of operation is the reverse of that of the generator mentioned before. The current to be measured is permitted to flow through the meter coil. This causes magnetic flux lines (magnetic lines of force) to be set up about the coil. These interact with the flux lines of the permanent magnet as shown in Fig. 1-3. This interaction produces a force that tends to align the magnetic field of the coil with that of the permanent magnet. This causes the coil to rotate. The magnitude of the force is proportional to the amount of current flowing in the coil. As the coil rotates, its movement is opposed by

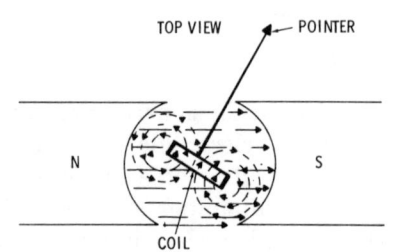

Fig. 1-3. Interaction between magnetic fields in a meter movement.

hairsprings whose operation is proportional to the amount of rotation. When the rotating force due to current flow and the opposing force due to the springs are equal, the coil comes to rest. As shown, a pointer is attached to the moving coil and moves across the calibrated meter face as the coil moves. The amount that the coil moves depends on the magnitude of the current. This is the most widely used type of meter movement. It is known by three names—permanent-magnet, moving-coil; d'Arsonval; or Weston movement.

A meter movement of this type can be made very sensitive, and instruments that deflect full scale with only ten microamperes (.00001 ampere) of applied current are available. Meter movements for the usual vom have sensitivities that fall in the ranges of 0-40 microamperes to 0-1 milliamperes (0-.001 ampere).

This so-called *sensitivity* of the meter movement represents the lowest scale of current measurement. However, the average vom is capable of measuring maximum currents of from several hundred milliamperes to as high as 10 or more amperes. This is accomplished by shunting, or diverting, most of the current through an associated calibrated resistance. Only a small percentage of the applied current then actually flows through the coil of the meter movement. These resistors are switched in and out of the circuit when you use the range switch of the vom.

The above arrangement often causes improper meter usage. If the meter is set on too low a scale in comparison to the current to be measured, a dangerously high current can pass through the meter coil. This excessive current can damage the coil assembly or bend the delicate pointer.

As you know, when a circuit is closed across a source of voltage, the electromotive force causes a current flow in that circuit. With the use of suitable calibrated resistors, an amount of current, that depends on the voltage that is connected across the meter terminals, can be made to pass through the meter movement. One needs only to calibrate the scale to measure voltage instead of current in this case. Again it is important that you do not apply a voltage higher than the maximum scale reading to which the meter has been set. If you do so, a higher than normal current is made to flow through the coil of the meter and damage it.

One or two batteries are a part of a vom. These batteries are used in the measurement of resistance. In association with calibrated resistances, the amount of current flow through the meter movement can be predicted in terms of an external resistance that is connected across the meter terminal. A resistance of a specific value attached across the meter terminals will cause a specific current flow through the meter movement. Thus it is possible to calibrate the meter scale in terms of resistance. The meter switch sets up the proper internal voltage and calibrated resistances to permit measurement within specific resistance ranges.

The Weston movement responds to a dc current and is basically a dc instrument. If ac electricity is applied, the meter movement will attempt to follow the changing alternations of the ac sine wave. However it cannot respond faithfully to the rapidly changing sine wave. As a result it will vibrate rapidly about a fixed position.

A dc movement can be used to measure ac electricity accurately by using a rectifier. A rectifier changes ac electricity to dc electricity. Hence it is possible to install a group of rectifiers inside the case of the vom. These are used whenever ac is to be measured. In association with calibrated resistances, the meter scale can be calibrated to measure ac current and voltage. All vom's are calibrated to read rms voltage; some, rms current.

EQUIPMENT AND CONSTRUCTION

The projects in this book are all constructed on two pegboards. Use machine screws and nuts to fasten the various parts to the pegboards. If the holes in the pegboard do not fall at exactly the positions needed, additional holes can be made with a small hand drill. Two separate pegboard arrangements are used throughout the book.

In the first several demonstrations, until you become familiar with circuit arrangements and drawings, the connections will be numbered and instructions on how to run specific wires among the various clips will be given. Later, you will be able to wire a circuit without being told where to run the specific wires. The materials needed for the science projects in Chapters 1 and 2 are given in the parts list of Table 1-3. You will

Table 1-3. Parts List for Chapters 1 and 2

Quantity	Description
1	VOM.
1	Pegboard, 12 in. by 8 in.
12	Fahnestock clips.
1	Spool of No. 20 push-back hookup wire.
1	Lantern battery, 6 V.
2	Filament transformer, 6.3 V, 1.2 A sec (Stancor P-6134 or equiv.).
1	Choke coil, 20 H, 900 Ω dc (Stancor C-1515 or equiv.).
1	Silicon rectifier, 1.0 A, 400 prv (HEP-157 or equiv.).
1	Transistor, general purpose (2N107, HEP-253, or equiv.).
1	Heat sink, clip-on (TO-5).
1	Switch, dpdt, knife type.
1	Switch, spst, knife or toggle type.
1	Line cord, ac.
1	Fuse holder, screw-terminal type (3AG).
1	Fuse (3AG, 2 A).
1	Neon bulb (NE-2).
1	Potentiometer, 25 kΩ.
2	Capacitor, 1000 μF, 15 V.
2	Capacitor, 100 μF, 12-25 V.
1	Capacitor, 25 μF, 12-25 V.
1	Capacitor, 0.1 μF.
1	Capacitor, 0.05 μF.
1	Resistor, 8200 Ω, ½ W.
1	Resistor, 2700 Ω, ½ W.
1	Resistor, 1800 Ω, ½ W.
2	Resistor, 1000 Ω, ½ W.
1	Resistor, 470 Ω, ½ W.
2	Resistor, 220 Ω, 2 W.
1	Resistor, 68 Ω, ½ W.
1	Resistor, 15 Ω, 5 W.

also need a number of 8-32 machine screws. These are usually sold in packages of 25 or more. Manufacturers' parts numbers are given for some parts to aid readers who wish to exactly duplicate the original demonstration. (Usually this is a wise decision.)

In the first project the pegboard will be set up to provide a choice of ac or dc electricity. The components are arranged as shown in Fig. 1-4.

A battery will be used as a source of dc voltage; a filament transformer will provide a source of ac power at 6.3 volts rms. The normal voltage for house wiring is about 110 volts, 60-Hz

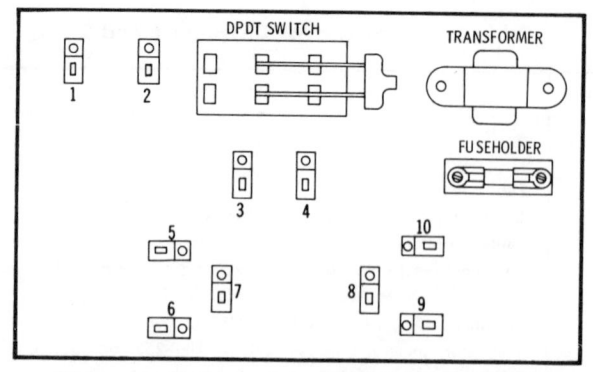

(A) Drawing showing parts location and clip identification.

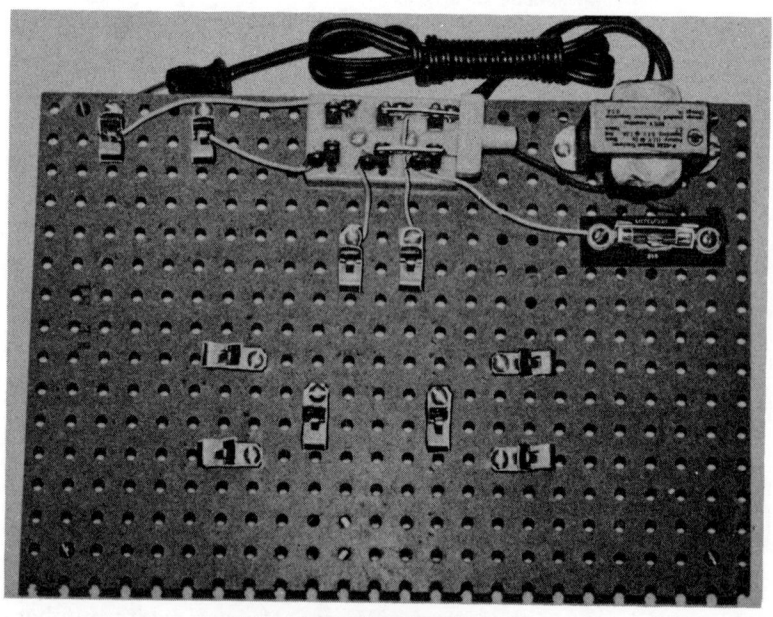

(B) Photograph of pegboard assembly.

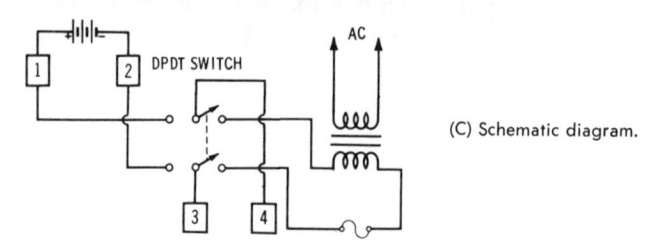

(C) Schematic diagram.

Fig. 1-4. Pegboard arrangement for obtaining ac or dc power.

Fig. 1-5. Method of connecting line cord to primary leads of transformer.

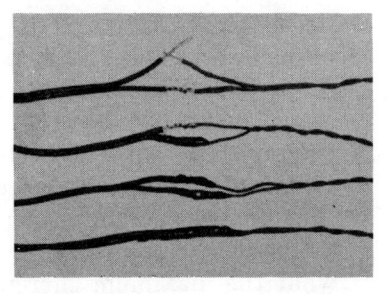

ac. To reduce the shock hazard, a transformer is connected between your power line and the pegboard circuit; it reduces your power line voltage to a safe 6.3 volts (If you are not sure what your house voltage is, check with your power company.)

The two primary leads of the filament transformer must be spliced and taped to the ac line cord (Fig. 1-5). The joined leads should be overlapped firmly as shown. The splice should be made by, or in the presence of, a person familiar with the wiring of 110-volt ac circuits. Each connection should be taped so that no part of either conductor is exposed. Neat and tight taping can be obtained by using only half-width electrical tape. This is done by cutting regular electricians' tape down the middle. You then have a length of half-width tape for each connection. A piece of full-width tape can then be used to bind the two segments together as shown.

Voltage can now be made available across the transformer secondary at any time the other end of the cord is inserted in an electric socket. No portion of the 110-volt ac wiring is exposed, and a possible shock hazard has been removed.

DEMONSTRATION 1
Using the VOM

The vom is a basic measuring instrument for electronics. It is used to measure voltage, current, and resistance, both ac and dc. A switching arrangement permits the choice of various voltage, current, and resistance ranges.

Introduction

To protect your vom several precautions should be taken in measuring voltage and current. These are:

1. Before applying power to any circuit or component to be measured, double-check to make certain that connections to the proper meter terminals have been made and that the meter switch has been set to measure the proper quantity (voltage, current, or resistance).
2. In measuring current, be certain that the meter scale is set to a higher maximum value than the highest current you might expect to find in the circuit. If you have no idea of what the maximum current could be, use the highest-current scale on the meter and gradually switch to lower and lower scales until you obtain a useful reading.
3. Always set the voltage-range setting of the meter to a higher range than the maximum voltage you might expect to find at the point of measurement. If you have no idea of what this voltage will be, use the maximum-voltage scale and decrease the voltage-range setting until you obtain a useful reading.
4. It is unwise to leave the meter set on the ohms position when it is not in use. In fact, it is a good habit to set any meter or instrument to its off position if it is not to be used for an extended period.

Procedure

1. Mount the double-pole, double-throw switch at the top center of the pegboard. A dc voltage source is connected to the left fixed terminals of the switch; an ac voltage source is connected to its right terminals. The two pole (center) terminals are connected to clips 3 and 4. Thus it is possible to switch between the ac and dc sources.
2. Connect clip 1 to the upper left terminal of the switch. Connect clip 2 to the lower left terminal of the switch. Connect the upper pole terminal to clip 4, the lower pole terminal to clip 3. Connect the plus (+) terminal of the battery to clip 1, the minus (−) terminal to clip 2.
3. Connect one lead from the secondary of the filament transformer to the upper right terminal of the switch. Connect the other lead of the transformer secondary to one side of the fuseholder. Connect the other terminal of the fuseholder to the lower right terminal of the switch. Insert the 2-ampere fuse into the holder. Keep the knife switch open.
4. Connect the common, or minus, side of the vom terminals

to clip 3. Usually the minus, or common, side of a vom is connected to the negative or ground side of a circuit or component. Notice that clip 3, when the switch is closed, is connected to the minus battery terminal. It is customary to use the black test lead of the vom in the minus or common circuit. Connect the positive (or +V) side of the vom to clip 4 by means of the red test lead.

5. Set the range switch of the vom on the dc 0 to 10- or 0 to 12-volt scale (use the lowest-voltage scale that permits a reading in excess of 6 volts). Close the knife switch to the battery side and check the reading.

   Momentarily reverse the vom test leads. What happens to the meter deflection? Restore the meter leads to the correct position. This check shows that in measuring the dc voltage it is necessary to use the correct meter polarity when connecting the test leads into the circuit. Always set your vom to the proper scale before making a measurement.

6. Throw the knife switch from the dc to the ac side. Observe the meter deflection. The basic movement of a vom reads dc electricity. To make an ac measurement, it is necessary to set the range switch to the ac-voltage position. Open the knife switch.

7. Set the vom to the ac 0 to 10-volt scale. Close the knife switch to the ac side. Observe the reading. In most cases the reading will be higher than 6.3 volts. An exact 6.3-volt reading occurs only when the primary ac voltage is of a specific value (usually 117 volts ac). If the primary voltage is other than this value, the secondary voltage will differ from 6.3 volts. The load placed on the secondary also influences the secondary voltage. Under rated load, the secondary voltage of the transformer will usually be about 6.3 volts.

8. Interconnect the six Fahnestock clips on the pegboard as shown in Fig. 1-6. These will be used for various circuit arrangements of resistors, capacitors, and inductors. Connect a wire jumper between clips 3 and 5. Connect a second wire between clips 6 and 7, and still another between clips 4 and 8.

9. Before connecting any components into the circuit, the vom is used to measure the value of several resistors used in the steps that follow. Set the ohmmeter to whatever

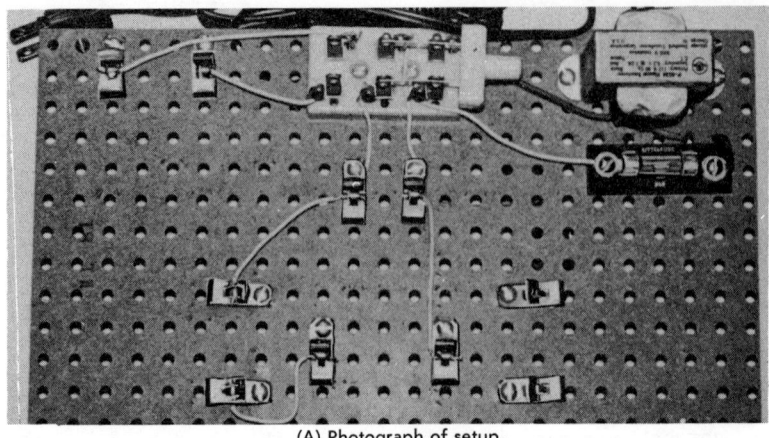

(A) Photograph of setup.

(B) Schematic diagram.

RESISTORS TO BE CONNECTED HERE

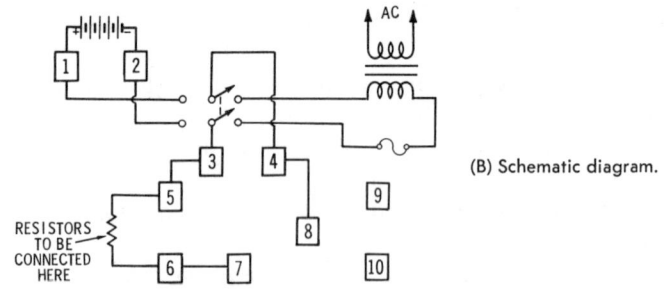

Fig. 1-6. Wiring of pegboard for Demonstration 1.

resistance scale is available between 0 to 2000 and 0 to 5000 ohms. Connect the ohmmeter leads to the two pigtail leads of a 1000-ohm, ½-watt resistor. Record the meter reading. The reading will not necessarily be exactly 1000 ohms. Resistors are rated in percentages. For example, if the percentage rating of the resistor is 10%, the actual resistance will fall somewhere between 900 and 1100 ohms.

Connect two 1000-ohm resistors in series (Fig. 1-7). Record the ohmmeter reading when it is connected between the outside leads. Connect the two resistors in parallel and take a similar ohmmeter reading.

When two like resistors are connected in series, the net resistance is double the value of one of the resistors. When two like resistors are connected in parallel, the net resistance is half the value of a single resistor.

10. Connect one of the 1000-ohm resistors in series with the 1800-ohm, ½-watt resistor. Record the ohmmeter read-

(A) Series connection.   (B) Parallel connection.

Fig. 1-7. Simple combinations of two resistors.

ing. Again the net resistance is the sum of the two individual resistors. In fact, any number of resistors can be added in series and the total resistance will be the sum of the individual resistors as stated in this simple formula:

$$R_T = R_1 + R_2 + R_3 + R_4 + \ldots$$

where,
R_T is the total resistance,
R_1, R_2, etc., are the individual resistances.

Connect the 1000-ohm and 1800-ohm resistors in parallel. Record the ohmmeter reading. The net resistance is lower in value than the lowest individual resistor of the pair.

The formula for the total resistance of a parallel combination of resistances is:

$$\frac{1}{R_T} = \frac{1}{R_1} + \frac{1}{R_2} + \frac{1}{R_3} + \frac{1}{R_4} + \ldots$$

where,
R_T is the total resistance,
R_1, R_2 etc., are the individual resistances.

In any parallel grouping of resistors, the net resistance is always lower than the smallest-value resistor of the group.

When there are just two resistors in parallel the following simple formula can be used. This is called a product-over-sum equation.

$$R_T = \frac{R_1 R_2}{R_1 + R_2}$$

In the specific case just mentioned:

$$R_T = \frac{1000 \times 1800}{1000 + 1800} = 643 \text{ ohms}$$

Notice that the calculation matches the ohmmeter reading for the parallel combination of the 1000-ohm and 1800-ohm resistors.

11. Insert a 1000-ohm resistor between clips 5 and 6. Place a short-circuit jumper between clips 7 and 8. Close the knife switch to the dc side. Measure and record the voltage across the resistor.

12. Remove the short circuit from between clips 7 and 8. Connect the common side of the vom to clip 7. Connect the other current terminal of the meter to clip 8. Set the vom to its highest current scale. Throw the knife switch to the dc side. Decrease the current-scale setting of the meter until a usable deflection of the meter is obtained (0 to 10 or 0 to 12 milliampere scale). Record the meter reading.

Use Ohm's law to calculate the current flow through the resistor. How does this compare to the meter reading?

$$I = \frac{E}{R}$$

$$I = \frac{6}{1000} = 0.006 \text{ ampere, or 6 milliamperes}$$

13. Connect two 1000-ohm resistors in series between terminals 5 and 6. Repeat the foregoing procedure.

$$I = \frac{E}{R_1 + R_2}$$

$$I = \frac{6}{1000 + 1000} = 3 \text{ milliamperes}$$

14. Connect two 1000-ohm resistors in parallel between terminals 5 and 6. Repeat the foregoing procedure. Calculate the current flow using Ohm's law.

$$I = \frac{E}{\frac{R_1 R_2}{R_1 + R_2}}$$

$$I = \frac{6}{\frac{1000 \times 1000}{1000 + 1000}} = 12 \text{ milliamperes}$$

15. Rearrange the circuit as shown in Fig. 1-8. Connect one 1000-ohm resistor between clips 3 and 5. Connect the

second 1000-ohm resistor between clips 3 and 6. Connect jumpers between clips 4 and 8, 7 and 8, and 5 and 7.

Insert the vom between clips 6 and 7. Close the knife switch to the dc side and record the current reading.

Open the switch and connect a jumper between clips 6 and 7. Remove the jumper between clips 5 and 7, and connect the vom in its place. Close the knife switch to the dc side and record the current reading.

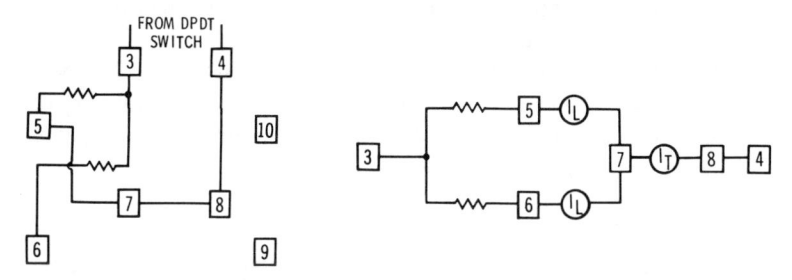

(A) Diagram of pegboard layout.　　(B) Diagram showing three metering points.

Fig. 1-8. Arrangement for measuring currents in a parallel circuit.

Open the knife switch. Place jumpers between clips 5 and 7 and between clips 6 and 7. Remove the jumper from between clips 7 and 8 and insert the current meter in its place. Close the knife switch and record the meter reading.

Note that the first two meter readings were equal and that the sum of these two readings equals the third reading. This step shows how the current divides in two paths, or legs, when resistors are connected in parallel. Use Ohm's law to calculate the current flow as follows:

$$\text{Current in each leg} = \frac{E}{R}$$

$$= \frac{6}{1000} = 6 \text{ milliamperes}$$

$$\text{Total current} = \frac{E}{\frac{R_1 R_2}{R_1 + R_2}}$$

$$= \frac{6}{\frac{1000 \times 1000}{1000 + 1000}} = 12 \text{ milliamperes}$$

16. Substitute the 1800-ohm resistor between clips 3 and 5 in place of the 1000-ohm resistor. Repeat procedure No. 15. Observe now that the two branch currents are unequal but that their sum still equals the total current. Use Ohm's law to verify the meter readings. *In a parallel circuit the sum of the leg currents equals the total current.*
17. Rearrange the circuit as shown in Fig. 1-9. Connect a 1000-ohm resistor between clips 5 and 6 and a second 1000-ohm resistor between clips 6 and 7. Connect a jumper between clips 7 and 4.
18. Close the knife switch to the dc side. Measure the voltage drop between clips 3 and 4, between clips 5 and 6, and between clips 6 and 7. Notice that the voltage drops across the two resistors are equal and their sum is equal to the voltage between clips 3 and 4. *In a series circuit the sum of the voltage drops equals the supply voltage.*
19. Calculate the series current by determining the current flow through one of the resistors. (In a series circuit the current through all parts of the circuit is the same.)

$$I_T = \frac{3}{1000} = 3 \text{ milliamperes}$$

20. Throw the knife switch to the ac side and make the same set of measurements as in Step 18. Be certain to set your vom to the ac voltage scale. Note that the operation of the circuit is identical. Although the usual vom does not permit the measurement of ac current, the current can be determined by measuring the ac voltage drop across either series resistor and using Ohm's law.
21. Remove the 1000-ohm resistor from between clips 6 and 7 and substitute the 1800-ohm resistor. Repeat the preceding three steps. Note that the voltage drops across the two resistors are not the same; however, the sum still equals the supply voltage.

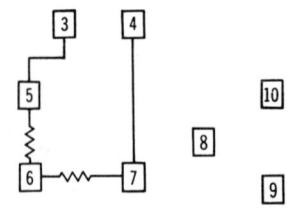

Fig. 1-9. Arrangement for measuring series voltage drops.

DEMONSTRATION 2
Inductors and Transformers

Whenever electrons are in motion, there is a *magnetic field* formed at right angles to the direction of current flow as shown in Fig. 1-10A. These lines are invisible, but their presence can be detected with a magnetic compass. This demonstration will show how the magnetic field is used in electronic circuits.

Introduction

The magnetic field strength, or *flux,* is strong near the conductor and decreases very quickly with distance from the

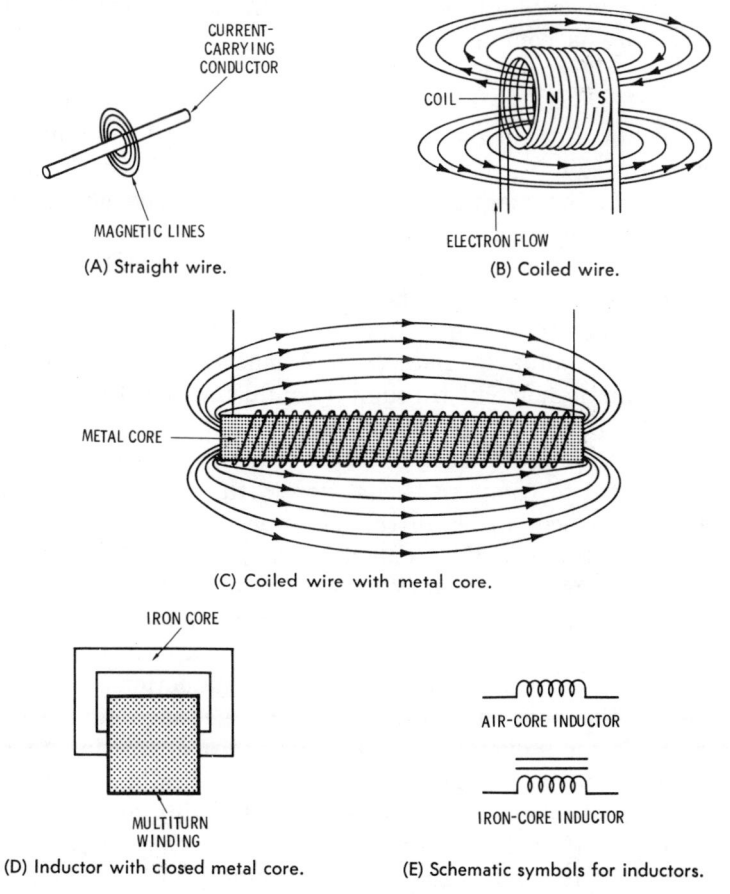

(A) Straight wire.

(B) Coiled wire.

(C) Coiled wire with metal core.

(D) Inductor with closed metal core.

(E) Schematic symbols for inductors.

Fig. 1-10. Inductors.

conductor. The higher the current flow, the stronger is the magnetic field. When the current flow in the conductor is reversed, the magnetic lines rotate in the opposite direction. It is important to know then that the strength of the magnetic field depends on the amount of current flow, while the direction of the magnetic lines depends on the direction of current flow.

If, instead of being straight, a conductor is concentrated in a small area by winding it in the form of a coil, it is possible to generate a strong, concentrated magnetic field (Fig. 1-10B).

If a piece of magnetic metal is placed in the center of the coil, there will be even less opposition to the magnetic flow, and an even stronger field is formed (Fig. 1-10C). Such a core is said to have a high *permeability*. The field is even stronger if the core is completely closed (Fig. 1-10D); that is, if it forms a continuous magnetic path.

When such an inductor is placed in a dc circuit, the magnetic field builds up and becomes stationary with current flow; it remains so until the current is changed or turned off. In the case of ac, the magnetic field changes continuously because of the changing current. The magnetic field surrounding a conductor in an ac circuit follows the sine-wave variations of the current passing through the winding.

This leads to another important electrical characteristic called *electromagnetic induction*. Whenever the lines of force of a magnetic field cut through a conductor, a voltage, or difference of potential, appears between the terminals of the conductor. The action is said to *induce* a voltage in the conductor. It is helpful to compare this action with that of the ac generator described earlier in this chapter.

In the latter case, the conductor was rotated while the magnetic field remained fixed. Whenever the magnetic field is stationary and the conductor moving, or the conductor stationary and the magnetic field changing, a voltage appears across the conductor as a result of electromagnetic induction.

One of the most unusual things about the magnetic field that surrounds an inductor is that it provides opposition to any change in the current flowing through the winding. When current begins to flow in an inductor, the changing magnetic lines induce an emf in the winding. The polarity of this induced emf is such that it opposes the source voltage. Hence, the inductor opposes the flow of current in the circuit (Fig. 1-11). However such opposition is not complete, and the current continues to flow and build up.

In a dc circuit, the current gradually builds up and the magnetic lines of force change less rapidly, the current finally reaching its maximum as determined only by the resistance of the coil and its external circuit. Once the current has built up to a maximum in a dc circuit it continues to flow unchanged as long as the voltage is applied. Because there is no change in current, the magnetic field is also constant. Actually the electrical energy has been converted to a magnetic field which persists so long as the current flows.

Fig. 1-11. Principle of self-induction.

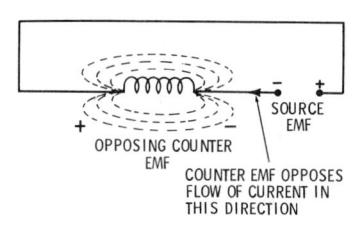

OPPOSING COUNTER EMF

SOURCE EMF

COUNTER EMF OPPOSES FLOW OF CURRENT IN THIS DIRECTION

The action is different in an ac circuit. Recall that an ac current is changing continuously. Hence there is a corresponding change in the magnetic lines of force. As a result, there is a continuous opposition to the flow of ac current in an inductor.

The higher the so-called *inductance* of the coil, the greater is the opposition to current flow. Inductance is measured in *henrys* and depends on the number of turns, the type of coil winding, and the permeability of the core. The actual opposition to ac current flow is called *inductive reactance,* which is measured in ohms.

One other factor influences the current flow in an inductor. This is the *frequency* of the applied current. The higher the frequency of an applied current the faster is the rate of change of the magnetic lines. When the magnetic lines cut the turns of the coil at a rapid rate, they induce a higher emf across the winding. This means there is even more opposition to the flow of current. Thus the inductive reactance of an inductor increases with frequency.

The above relations tell us that the inductive reactance of an inductor varies directly with the inductance and the frequency. Stated as a formula:

$$X_L = 2\pi f L$$

where,

X_L is the inductive reactance in ohms,
L is the inductance in henrys,
f is the frequency in hertz,
π is 3.14.

There are three basic types of magnetic induction. An emf is induced in a conductor when that conductor is moved in a magnetic field or when magnetic lines of force cut through a fixed conductor. You also learned that an emf can be induced in an inductor because its own magnetic lines of force cut through its windings. This is called *self-induction*. The third type of induction is called *mutual induction*.

When a current flows in the winding of an inductor, a magnetic field is formed. This field expands outward from the inductor according to the inductance and the variations of the current flow. The magnetic field expands and contracts with the changing current in the winding. If you now place a second inductor in the magnetic field that surrounds the first, the moving lines induce a voltage across the second inductor. This is mutual induction. It occurs only when the magnetic lines of force about the first inductor are changing and are cutting through the windings of the second inductor.

(A) Construction of iron-core transformer.

(B) Symbol for air-core transformer.

(C) Symbol for iron-core transformer.

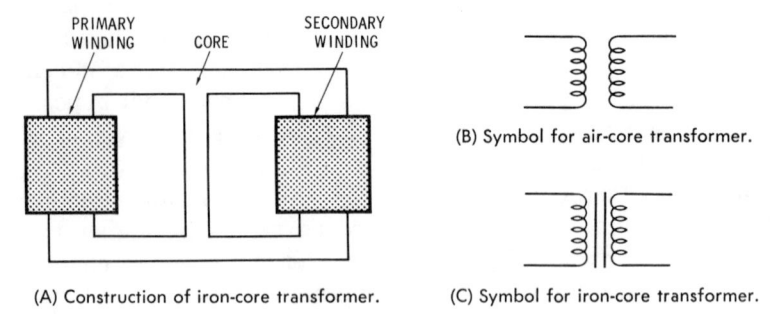

Fig. 1-12. Simple transformer.

Two such windings coupled near to each other are said to be *inductively coupled*. Such an arrangement is called a *transformer* (Fig. 1-12). The inductor to which the electrical energy is being applied in the form of a changing current flow is called the *primary*. The other winding is the *secondary*.

The amount of voltage induced in the secondary from the primary is related to the *turns ratio*. This is the ratio of the number of primary and secondary turns. A transformer is said to have a step-up ratio when there are more secondary turns than primary turns. Conversely, if the transformer has fewer turns on the secondary than the primary, it is called a step-down type. In the project, a step-down transformer is used to decrease the 110-volt ac voltage from a power line to 6.3 volts.

Procedure

1. In this demonstration you will work with a series resistor-inductor combination. It will be connected as shown in Fig. 1-13. Before connecting the choke inductor between clips 9 and 10, measure and record the dc resistance of the primary winding. Insert a 1000-ohm, ½-watt resistor between clips 7 and 8. Place jumpers between clips 3 and 7 and clips 10 and 4.
2. Throw the knife switch to the dc side and measure the dc voltage across the inductor and across the 1000-ohm resistor. Note that the two values are comparable, the division being determined by the resistance values.
3. Remove the jumper from between clips 3 and 7. Insert the vom between these two points. Measure the dc current flow in the circuit.
4. Use Ohm's law to prove that this amount of current should flow. Note that the inductance of the inductor does not impede the flow of dc current. The dc flow in the inductor is limited only by the dc resistance of its winding.

$$I = \frac{E}{R_1 + R_W}$$

$$I = \frac{6}{1000 + 900} = 3.15 \text{ milliamperes}$$

where,
 I is the current,
 E is the applied voltage,
 R_1 is the resistance of the resistor,
 R_W is the resistance of the choke.

5. Remove the vom from the circuit and replace the jumper between clips 3 and 7. Throw the knife switch to the ac

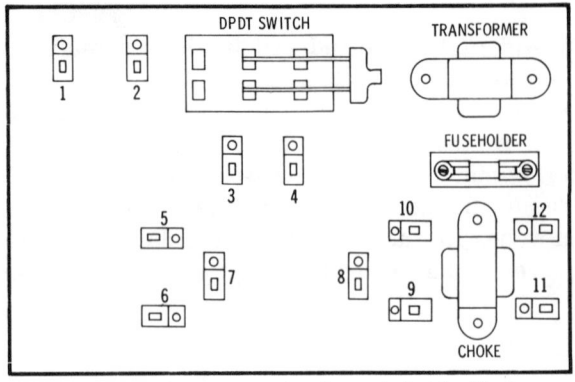

(A) Drawing showing parts location and clip identification.

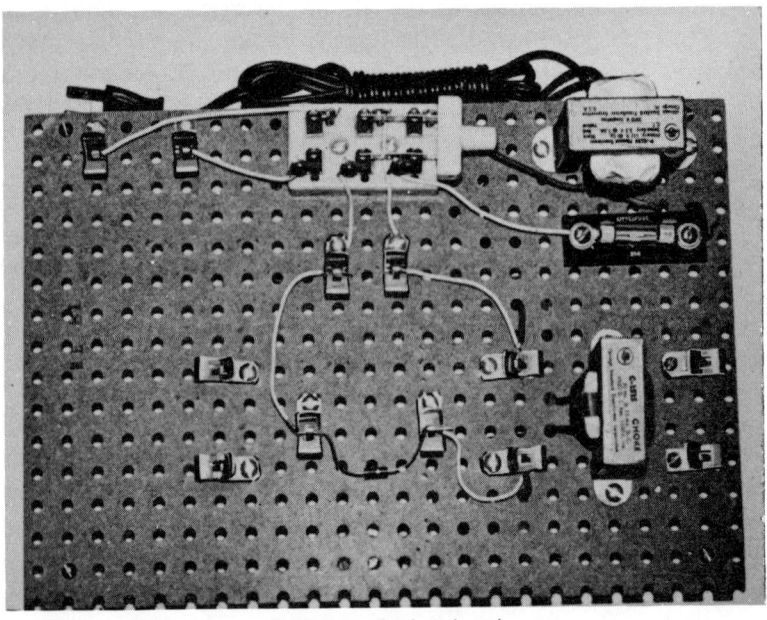

(B) Photograph of pegboard.

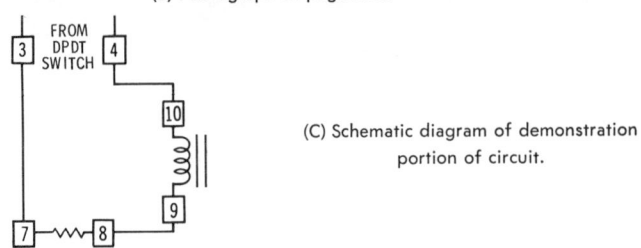

(C) Schematic diagram of demonstration portion of circuit.

Fig. 1-13. Arrangement for inductor demonstration.

side. Measure the voltage drop across the inductor and the resistor. Make a comparison with those readings obtained when using a dc supply voltage. What has happened?

6. What is the ac current flow in the circuit? Recall that ac current flow in the series circuit can be calculated by using Ohm's law and the ac voltage measurement across the resistor. Compare this with the current flow when the dc voltage was applied to the same circuit. Notice how much the inductor limits the flow of ac current in comparison with Step 4.

$$I = \frac{E_R}{R_1} = \frac{.35}{1000} = 0.35 \text{ milliampere}$$

where,
I is the current,
E_R is the voltage across the resistor,
R_1 is the resistance of the resistor.

7. Remove the inductor and mount a filament transformer at the same position. Connect the 6.3-volt side (primary) between clips 9 and 10. Connect the outside leads of the 117-volt side (secondary) between clips 11 and 12. Use the circuit arrangement shown in Fig. 1-14.

8. Insert a 68-ohm, ½-watt resistor between clips 7 and 8. Connect a neon bulb between clips 11 and 12.

9. Close the knife switch to the dc side. Measure the dc voltage drop across the primary of the transformer and across the resistor. Record these readings. Remove the jumper from between clips 3 and 7. Connect the vom in the same position. Measure the dc current flow. Remember to set the vom on a high enough current range. Note that a substantial dc current flows.

10. Disconnect the meter and replace the jumper between clips 3 and 7. Set the knife switch to the ac side. Measure the voltage drop across the resistor and across the primary of the transformer. Record these readings. Measure the voltage drop across the secondary of the transformer. Note that there is now a greater voltage drop across the transformer primary because of the influence of the reactance on the ac current flow.

11. Replace the resistor between clips 7 and 8 with a jumper. Set the knife switch to the dc position. Although there is

now a substantial primary current flow, no voltage is developed across the output side of the transformer. Observe the neon bulb across the output whenever the knife switch is opened. Do not permit current flow longer than necessary because of the heavy drain on the battery.

It is important to realize that a transformer does not operate when there is a dc current flow. The neon bulb glows momentarily when the knife switch is opened because of the collapse of the magnetic field about the transformer. In cutting through the secondary winding these collapsing magnetic lines induce a high enough voltage (in excess of 63 volts) to operate the neon bulb.

12. Throw the knife switch to the ac position. Notice that the neon bulb now glows brightly. Measure the voltage drop across primary and secondary. Why is the secondary voltage so high? What is the step-up ratio of the transformer?

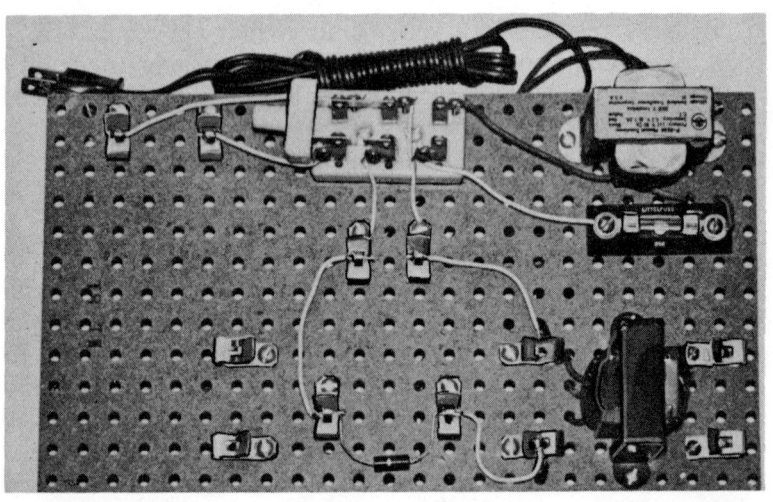

(A) Photograph of pegboard.

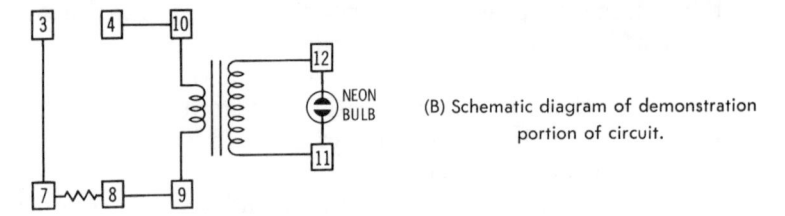

(B) Schematic diagram of demonstration portion of circuit.

Fig. 1-14. Arrangement for transformer demonstration.

$$\text{Step-Up Ratio} = \frac{E_s}{E_p}$$

where,
E_s is the secondary voltage,
E_p is the primary voltage.

DEMONSTRATION 3
Capacitors and Resonant Circuits

Lines of stress are also present as a result of the voltage, or potential difference, between two bodies. These are called *electric lines* of force as opposed to *magnetic lines* of force. It is the purpose of this demonstration to show some of the practical effects of these lines of force.

Introduction

If one of two bodies has an excess of electrons, it is said to have a negative charge. The other, because it has a deficiency of electrons, is said to be charged positively. Stress lines then extend between the two bodies.

Two conductive metal plates can be charged by connecting them between a source of voltage. If a 3-volt battery is connected between the plates as in Fig. 1-15, electrons will flow out of the negative battery terminal to the left plate. However, electrons cannot move from the left to the right plate because the plates do not touch each other. Electrostatic lines of force are set up between the two plates. This stress forces electrons

Fig. 1-15. Operation of basic capacitor.

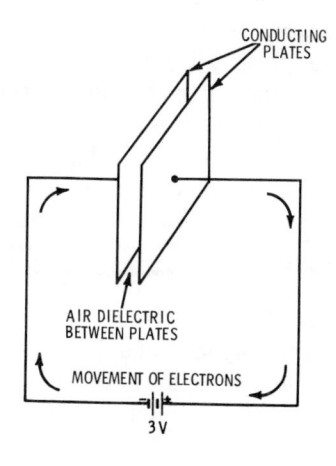

from the right plate to flow down through the other conductor to the positive terminal of the battery. The net result is that there is an excess of electrons on the left plate and a deficiency of electrons on the right plate.

Electrons build up on the left plate and leave the right plate until the difference of potential between plates is the same as the battery voltage. Then there is no further flow of current. The electrical energy is stored in electric lines of force between the plates.

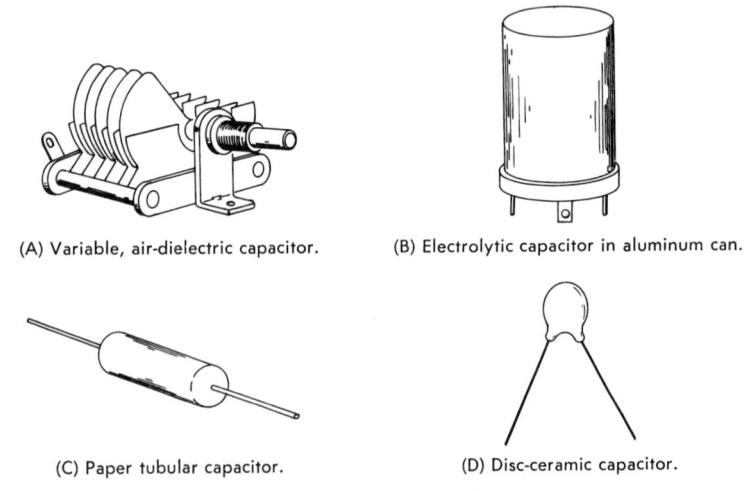

(A) Variable, air-dielectric capacitor.

(B) Electrolytic capacitor in aluminum can.

(C) Paper tubular capacitor.

(D) Disc-ceramic capacitor.

Fig. 1-16. Some typical capacitors.

If the battery voltage is changed, there will be electron motion again. This motion of electrons will continue long enough to bring the charge between the two plates to the same difference of potential as the new voltage.

A device consisting of two such conducting plates separated by an insulator is called a *capacitor*. Its ability to store an electric charge is called *capacitance*. The unit of capacitance is the *farad*. Capacitance depends on the size and shape of the conductors and the spacing between the conductors. Capacitors, like inductors, come in a variety of sizes, shapes, and construction (Fig. 1-16).

The capacitance is also determined by the characteristics of the insulator between the two conductors. This insulating material is called the *dielectric*. Its relative ability to store electric energy is called its *dielectric constant*. The higher the dielectric constant of the insulator the more readily it stores a large amount of electrostatic energy.

In an ac circuit the applied voltage changes continuously. Thus a steady charge never builds up on the capacitor. Electrons are being continuously moved on and off the two plates. Thus current flows in and out of the capacitor all the time following the voltage variations of the ac source.

The opposition that a capacitor offers to the flow of ac current depends on its capacitance. The opposition to the flow of ac is again called reactance. This time it is called *capacitive reactance*. The unit of capacitive reactance is also the ohm.

The greater the capacitance of a capacitor, the greater the electron flow into and out of the plates. Consequently a capacitor with a large capacitance offers a lesser opposition to the flow of ac than one of lower capacitance.

The reactance also depends on the frequency. The higher the frequency of the applied voltage, the faster the capacitor must charge and discharge, and therefore the rate of current flow into and out of the capacitor must be higher. In effect there is less opposition to current flow when the voltage is changed at a rapid rate (high frequency).

The preceding relationships tell us that the capacitive reactance varies inversely with the frequency and the capacitance. Stated as a formula:

$$X_C = \frac{1}{2\pi f C}$$

where,
X_C is the capacitive reactance in ohms,
C is the capacitance in farads,
f is the frequency in hertz,
π is 3.14.

One of the unusual conditions that arises when capacitors and inductors are used in the same circuit is a phenomenon called *resonance*. Although capacitive reactance and inductive reactance are given in ohms, the two reactances are not the same. Inductive reactance is usually referred to as *positive reactance*; capacitive reactance is called *negative reactance*. When the two are put together, one reactance subtracts from the other.

If a capacitor and inductor in a circuit have exactly the same reactance, one would subtract from the other to give zero reactance. This means that there would be no reactive opposition to the flow of ac current. Since inductive reactance and capacitive reactance both change with frequency, this condition can only happen at one specific frequency. Because resonance occurs at only one frequency, appropriate combinations of inductors and capacitors can be used to emphasize or de-emphasize a signal of a specific frequency. For this reason resonant circuits are used widely in radio equipment.

The frequency at which inductive reactance and capacitive reactance are the same is called the *resonant frequency*. The fundamental resonance formula is:

$$f_r = \frac{1}{2\pi \sqrt{LC}}$$

where,
f_r is the resonant frequency,
L is the inductance in henrys,
C is the capacitance in farads,
π is 3.14.

Procedure

1. Rearrange the pegboard as in Fig. 1-17. Connect the 25-microfarad (μF) capacitor between clips 5 and 6. Connect the 68-ohm resistor between clips 7 and 8.
2. Set the knife switch to the dc side. Measure the dc voltage drop across the capacitor and across the resistor. Note that the capacitor charges to the supply voltage and that there is no voltage drop across the resistor.
3. Remove the jumper from between clips 3 and 5. Connect the vom between these two clips. Set the vom to its most sensitive dc current scale. Close the switch to the battery side and notice the meter action. When the switch is first closed there will be a momentary flow of current. When the capacitor is charged, current flow will stop as indicated by the fall-back of the meter reading to zero. Remove the vom from this position and reconnect the jumper between clips 3 and 5.
4. Throw the switch to the ac side. Measure the ac voltage across the capacitor and the resistor. Use Ohm's law to calculate the current flow.

$$I = \frac{E_R}{R} = \frac{4.4}{68} = 64.7 \text{ milliamperes}$$

5. Replace the 25-μF capacitor with a 100-μF capacitor. Throw the switch to the ac side again. Measure the ac voltage drop across the capacitor and the resistor. Compare

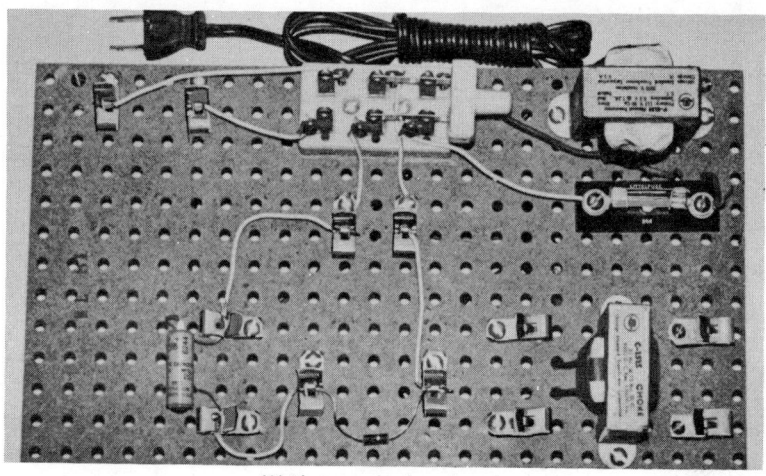

(A) Photograph of pegboard.

(B) Schematic diagram of demonstration portion of circuit.

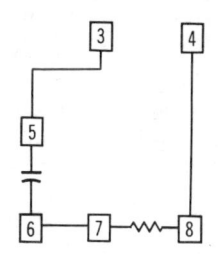

Fig. 1-17. Arrangement for capacitor demonstration.

it with the previous values. Again calculate the ac current flow. Note that the current flow is greater because of the lower reactance of the larger capacitor.

$$I_T = \frac{E_R}{R} = \frac{7.2}{68} = 106 \text{ milliamperes}$$

6. In this part of the demonstration both a capacitor and inductor will be used. Rearrange the pegboard according to Fig. 1-18. Insert the 0.1-μF capacitor between clips 5 and

6. Connect the 20-henry choke between clips 9 and 10. Insert a 1000-ohm resistor between clips 7 and 8.
7. Connect the vom across the resistor. Set the vom to its most sensitive dc scale. Throw the knife switch to the dc side. Notice that there is no meter reading, indicating that there is no current flow because of the blocking action of the capacitor.

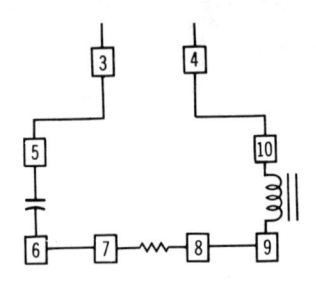

Fig. 1-18. Schematic diagram of additional portion of circuit for resonance demonstration.

8. Open the knife switch. Set the vom to its most sensitive ac scale. Throw the knife switch to the ac side. Record the meter reading. From the reading, calculate the ac current flow.

$$I_{RES} = \frac{E_R}{R} = \frac{1.5}{1000} = 1.5 \text{ milliamperes}$$

where,
 I_{RES} is the current at resonance,
 E_R is the voltage across the resistor,
 R is the resistance of the resistor.

9. Momentarily short out the capacitor with a jumper. Record the meter reading and calculate the ac current. Remove the short from across the capacitor and place it across the inductor. Record the meter reading and calculate the ac current flow.

$$I_1 = \frac{E_R}{R} = \frac{.3}{1000} = 0.3 \text{ milliampere}$$

$$I_2 = \frac{E_R}{R} = \frac{.35}{1000} = 0.35 \text{ milliampere}$$

where,
 I_1 is the current with the capacitor shorted,
 I_2 is the current with the inductor shorted.

Notice that there is more current flow when both capacitor and inductor are in the circuit. If only one is present in the circuit there is less current flow. These results show the cancellation effect of capacitive reactance and inductive reactance.
10. Remove the short from across the inductor. Set the vom to the 0 to 60 or 0 to 100 ac voltage scale. Measure the voltage across the capacitor. Measure the voltage across the inductor. Measure the voltage between clips 3 and 4.

    An unusual condition of resonance is demonstrated. The voltages across the capacitor and the inductor are greater than the supply voltage. Actually the voltage across the inductor is of one polarity and the voltage across the capacitor is of the other polarity at each instant. Thus there is a voltage-cancelling effect and the net voltage across the three components is equal to the supply voltage.

    This resonant condition shows how resonant circuits can be used to emphasize a signal having a particular frequency. Only at this one frequency is there such a high voltage and high series-current flow.
11. Reconnect the vom across the resistor. Set the vom once more to its most sensitive ac voltage scale. Place a 0.05-μF capacitor in parallel with the 0.1-μF capacitor. Notice that the current flow now decreases as indicated by the drop in voltage across the resistor.

    Measure the ac voltage across the capacitor using the appropriate ac voltage scale. Measure the ac voltage across the inductor. Now the inductor voltage is higher than the capacitor voltage because the capacitive reactance has been decreased below that of the inductive reactance.
12. Reconnect the vom across the resistor. Reset the meter to its most sensitive ac voltage scale. Remove the 0.1-μF capacitor, leaving only the 0.05-μF capacitor in the circuit. Note that the current flow as indicated by the lower voltage across the resistor is again lower than at resonance.

    Measure the ac voltage across the inductor using the appropriate meter scale. Measure the ac voltage across the capacitor. Now the capacitive voltage is greater than the inductive voltage because the decrease in capacitance results in a higher capacitive reactance than inductive reactance.

By increasing or decreasing the capacitance you have shifted the resonant frequency of the inductor-capacitor combination away from 60 Hz. Thus the reactances do not entirely cancel, and the uncancelled reactance limits the current flow. At resonance the two reactances are the same and cancel completely; the current is limited only by the resistance of the circuit, and a high current can flow.

2

Diode and Transistor

There are about ten basic electronic components. The maze of parts that make up an electronic unit are, in general, modifications or variations in the size of these basic building blocks. In the previous chapter, important parts such as resistors, capacitors, inductors, and transformers, were used in the project demonstrations. Two other fundamental components are the tube and the semiconductor device; a transistor is an example of the latter. Semiconductor devices are the subject of this chapter.

RECTIFIER OPERATION

A rectifier is a unidirectional device. It permits electric current to flow strongly in one direction through a circuit; little or no current in the opposite direction. There are three popular types of rectifiers used in the power circuits of electronic equipment. These are the vacuum-tube rectifier, the contact rectifier, and the semiconductor rectifier. The semiconductor type is used as a power rectifier for the various projects in this book.

A typical semiconductor silicon rectifier is shown in Fig. 2-1. A semiconductor material is defined as one that permits a controlled amount of current flow when the proper operating conditions are established. In the case of a silicon rectifier, there

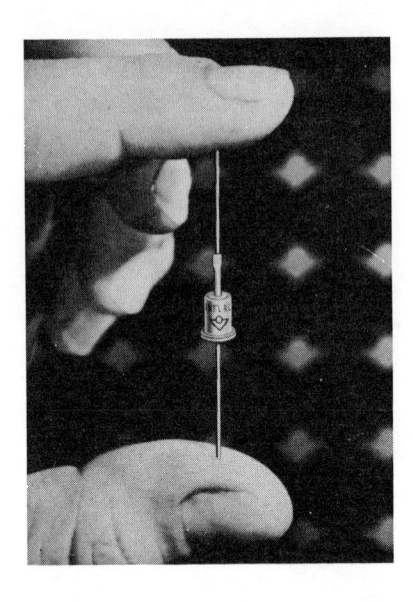

Fig. 2-1. Typical silicon rectifier.

is an electron current flow whenever a positive difference of potential, or voltage, is present between its anode and cathode. There is no significant current flow when the anode is negative with respect to the cathode. This fact makes it possible for a silicon rectifier and its associated circuit to convert ac electricity to dc electricity.

Two basic rectifier circuits are shown in Fig. 2-2. In Fig. 2-2A an ac voltage from the secondary of a transformer is applied to the rectifier. On the positive alternation of the sine wave, the anode is positive with respect to the cathode. Thus the rectifier permits current flow. As a result there is a current flow from the low end of the transformer secondary, through the load resistor R_L, and on through the rectifier to the other end of the secondary. On the negative alternation of the sine wave, the anode is negative with respect to the cathode, and there can be no current flow.

Although the current flows in only one direction, the actual magnitude of the current is not constant because the positive alternation of the sine wave starts from zero, swings up to maximum, and returns to zero again. Thus the current flow has the shape of the waveform shown in Fig. 2-2A. This type of waveform is called *pulsating dc*. To make such a waveform useful in most electronic circuits it has to be filtered as discussed in the next section. Since there is current flow through the output circuit only during one half-cycle of the applied sine-wave voltage, this circuit arrangement is called a half-wave rectifier.

The rectifier circuit of Fig. 2-2B is called a full-wave rectifier because current flows in the load for both polarities of the applied sine wave. Notice that the cathode side of each rectifier is returned through the load resistor to the center of the transformer secondary. On the positive alternation of the applied

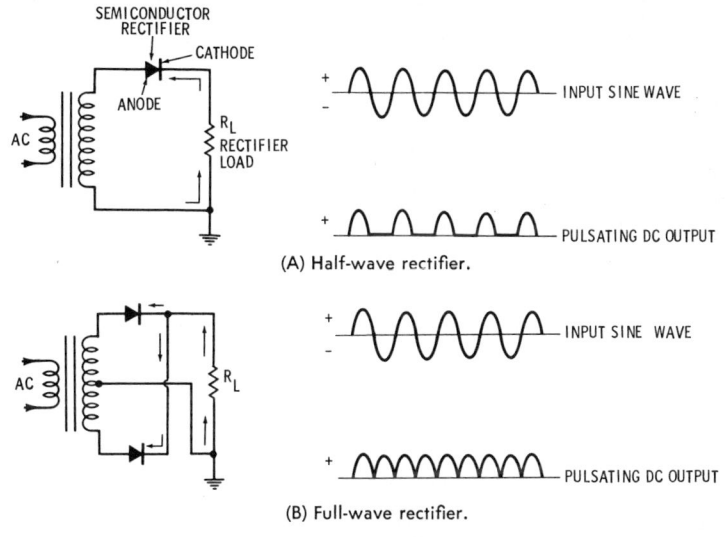

Fig. 2-2. Typical rectifier circuits.

sine wave, the anode of the top rectifier is made positive with respect to its cathode. Consequently, current flows through the load resistor for each positive alternation. During the positive alternation of the sine wave across the secondary, the voltage applied to the anode of the bottom diode is negative with respect to its cathode. Consequently, the bottom diode is not conducting.

On the negative alternation of the sine wave there is a negative voltage applied to the anode of the top diode and, therefore, it is nonconducting. Since the bottom diode is connected to the lower end of the transformer, its anode is actually positive with respect to its cathode during the negative alternation of the sine wave. Thus, the lower diode conducts and allows current to flow through the load resistor. Note that the arrangement of the circuit and the directions of conduction in the diodes are such that current always flows through the load in the same direction.

Since the two rectifier diodes conduct on opposite alternations of the sine wave, the waveform shown in Fig. 2-2B results. It is apparent that current flow is much more continuous than in the case of the half-wave rectifier. A full-wave rectifier provides a higher possible current flow, and it is less difficult to smooth out the pulsations in its output to a pure dc with a filter.

FILTERING

The purpose of the filter is to make a pure dc voltage out of pulsating-dc output of the rectifier. A filter uses either a capacitor-inductor combination or a capacitor-resistor combination to do this (Fig. 2-3). A capacitor opposes a change in

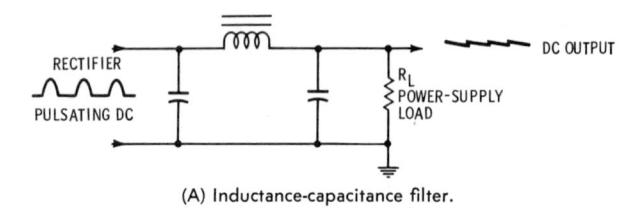

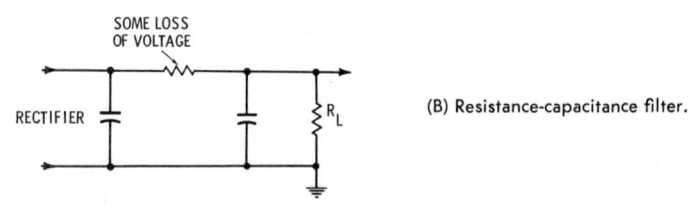

(A) Inductance-capacitance filter.

(B) Resistance-capacitance filter.

Fig. 2-3. Basic filter circuits.

voltage, and an inductor opposes a change in current. A resistor-capacitor combination also tends to smooth out any voltage variation. The higher the value of the capacitor and resistor the better is the filtering action. Too high a value of resistor, however, cuts down the amount of dc output voltage that can be obtained, particularly when a high current is drawn. In summary, the rectifier changes the ac electricity made available at the secondary of the transformer to pulsating dc electricity. A filter circuit smooths out the pulsations, or *ripple*, to obtain pure dc electricity comparable to that which can be obtained from a battery.

THE SEMICONDUCTOR JUNCTION

Diode and transistor operation can be better understood if one first considers the activity that takes place at the border or *junction* between the two segments of a semiconductor device. The type of impurity mixed in with a semiconductor material determines its polarity. One type of impurity provides an excess of electrons and freedom of motion for negative charges. This is called *n-type* semiconductor material. The impurity in other semiconductor material may be such that positive charges can move freely. These positive charges are often referred to as *holes* because they represent vacant electron positions in the atomic make-up of the material. This is called *p-type* semiconductor material.

Much happens when two such segments are joined together to form a junction. The motion of the charges depends on the polarity of any external voltage applied across the junction. When a negative voltage is applied to the n-type material and a positive voltage to the p-type material, there is a flow of current. The negative potential on the n-type material will repel the negative charges, driving the electrons toward the junction between the two segments. In a similar manner the positive potential on the p-type material will drive the positive particles, or holes, toward the junction. Consequently, there will be a free motion of charges across the junction, and a low-resistance conducting path results (Fig. 2-4A).

This is the basic principle of operation of the silicon rectifier. The p-type material is called the *anode*; the n-type material, the *cathode*. Whenever a voltage that is positive with respect to the cathode is applied to the anode, there is a flow of current in the external circuit. The junction is then said to be forward biased.

Fig. 2-4B shows the activity when the p-type material is made negative with respect to the n-type material. In this case the negative charges, or electrons, are drawn toward the positive terminal. In a similar manner the positive charges, or holes, are drawn toward the negative terminal. As a result the charges are removed from the junction, and there is no exchange of charges between the segments. Thus a continuous flow of current can not be established due to the very high resistance of the junction. In this case the junction is said to be reverse biased, or back biased. No current flows in the external circuit. In rectifier operation this condition exists during the negative alternation of the applied sine wave.

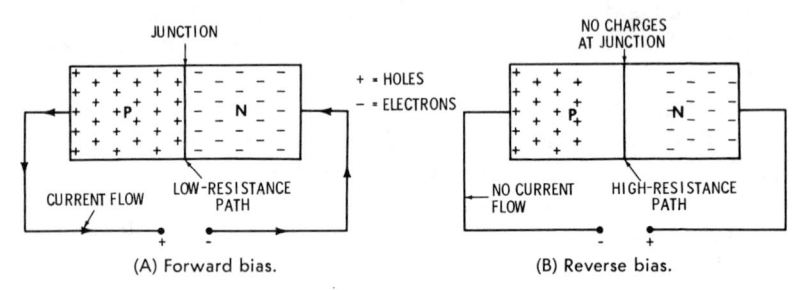

Fig. 2-4. Semiconductor diode operation.

The so-called *pn junction* just described has a low resistance path when it is forward-biased, permitting a high current flow. When it is reverse-biased, it has a high resistance; little or no current can flow.

THE BIPOLAR TRANSISTOR

The *bipolar* transistor can be used as a voltage or current amplifier. This is possible because a bipolar transistor has two junctions instead of one (Fig. 2-5). Note that the center segment is a different type of semiconductor than the two outer segments. In Fig. 2-5A the outer sections are n-type and the center section is p-type; this is referred to as an *npn* transistor. The center segment is called the *base*; the outer segments, *emitter* and *collector*. There are also *pnp* transistors (Fig. 2-5B).

The following four factors are important in understanding the operation of a transistor circuit:

1. In an operating circuit, the collector-base junction is reverse biased. Thus the collector-base junction has a high resistance.

2. When no voltage (zero bias) is applied to the base-emitter junction, there is no current flow through either junction.
3. If the base-emitter junction is reverse biased, it also has a high resistance. Again there is no current flow through either junction.
4. When the base-emitter junction is forward biased, as it is in normal operation, there is a flow of base-emitter current due to the low resistance of this junction. Activity inside the transistor is such that there is also a strong

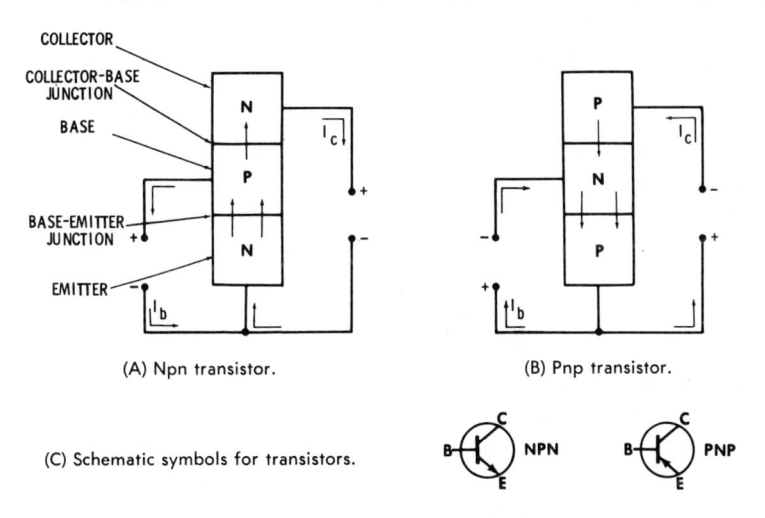

(A) Npn transistor.

(B) Pnp transistor.

(C) Schematic symbols for transistors.

Fig. 2-5. Transistor operation.

current flow through the collector-base junction. Normally this collector current flow is greater than the base current flow; the transistor stage is thus operating as a current amplifier.

Consider now the activity that takes place when, as in normal operation, the collector-base junction is reverse biased and the base-emitter junction is forward biased. A small amount of signal current variation in the base circuit causes a substantially larger variation in the collector current. This collector current flows in the load circuit connected between the collector and the emitter.

In Fig. 2-5A, the dc supply sources apply a reverse bias to the collector-base junction and a forward bias to the base-emit-

45

ter junction. The forward bias on the base-emitter junction forces positive and negative charges to the junction, and a base current (I_b) flows. As a result electrons move into the base element. Some of them are neutralized by the positive charges or holes in the base. However, more electrons move into the base than there are neutralizing holes. These are attracted toward the collector because of the positive potential of the collector. Thus there is a strong flow of electrons from the emitter into the collector through the base. This flow is much stronger than the base current.

A similar activity occurs in a pnp transistor. Pnp and npn transistors are said to have complementary characteristics. In the case of the pnp transistor, the forward biasing of the base-emitter junction causes a rush of positive charges across the base-emitter junction because of the low-resistance path. These are in part neutralized by the electrons of the base. However, the positive charges are attracted strongly by the negative potential of the collector. Thus they cross the base-emitter junction in quantity and move through the base across the collector-base junction into the collector region. Again there is a strong flow of collector current.

In the pnp circuit, current flow is in the opposite direction from that of the npn transistor. Note that a negative potential must be applied to the p-type collector material in order to reverse bias the collector-base junction. Therefore the electron current flow in the external circuit is in the opposite direction as compared to the npn circuit of Fig. 2-5A.

DC AND AC OPERATION

In the previous discussion, dc currents have been considered. The ratio of the dc collector current to the dc base current is referred to as the dc current gain, or *dc beta*. In the *common-emitter* circuit shown in Fig. 2-6A, the dc current gain is often given the symbol h_{FE}.

An applied audio or other signal causes the base current to vary about the base-bias current set by the dc base-bias voltage (Fig. 2-6B). This base-current variation will, in turn, cause an amplified change in the collector current. The ratio of the ac collector current to the ac base current is the *small-signal beta* of the transistor stage or:

$$h_{fe} = \frac{\Delta I_c}{\Delta I_b}$$

where,

h_{fe} is the small-signal current gain,
ΔI_c is a small change in collector current,
ΔI_b is the corresponding small change in base current.

The actual output voltage E_O results from the variations of the collector current in the collector load. The input base-current variation was caused by the input voltage E_{IN}. Hence the actual voltage gain of the transistor stage is the ratio of E_O/E_{IN}.

It is important to note the polarity of the input and output voltages. In Fig. 2-6, a pnp transistor is used; its base is negative with respect to the emitter, and its collector is negative with respect to the base. As shown in Fig. 2-7, a positive swing of the base voltage will decrease the base current. In turn the collector current will decrease. The decrease in collector current through the load resistor makes the output voltage swing negative. Conversely, the negative alternation of the input voltage will cause an increase in base current, an increase in collector current, and a positive swing of the output voltage. Input and output voltages are said to be of opposite polarity or *out of phase*.

As a matter of information, some typical phasings between sine waves are shown in Fig. 2-8. When the timing of the two waves is identical, they are said to be in phase. It is customary

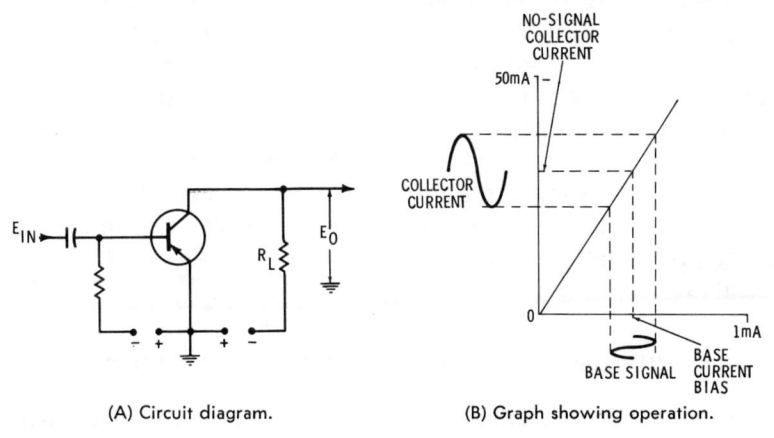

(A) Circuit diagram.

(B) Graph showing operation.

Fig. 2-6. Basic transistor-amplifier stage.

47

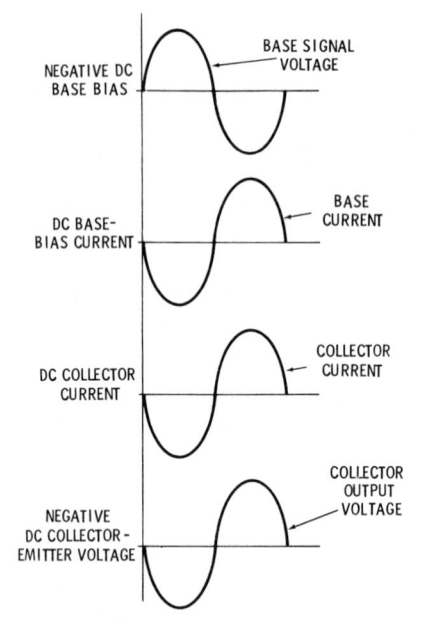

Fig. 2-7. Signal currents and voltages in pnp transistor-amplifier stage.

to subdivide a sine wave into degrees as we do a circle. A full sine wave thus has 360 degrees. This type of measurement is based on the relation of the sine wave to the rotation of the generator coil.

If a second wave doesn't begin its departure from zero until a half-cycle after the first one (180 degrees later), the two sine waves are said to have a phase difference of 180 degrees. This is the relationship between the input and output voltages in Fig. 2-7. Thus, out-of-phase, opposite-polarity, and 180-degree related sine waves are different names for the same condition.

Waves can be other than in phase or out of phase as shown in Figs. 2-8D and 2-8E. In Fig. 2-8D the second sine wave doesn't begin its rise from zero until the first one has reached a maximum, or 90 degrees later. The two sine waves are said to be 90-degree related. In Fig. 2-8E the two waves are 45-degree related.

THE UNIPOLAR OR FIELD-EFFECT TRANSISTOR

The field-effect transistor (FET) is a solid-state device with a very high input impedance and a high output impedance. A

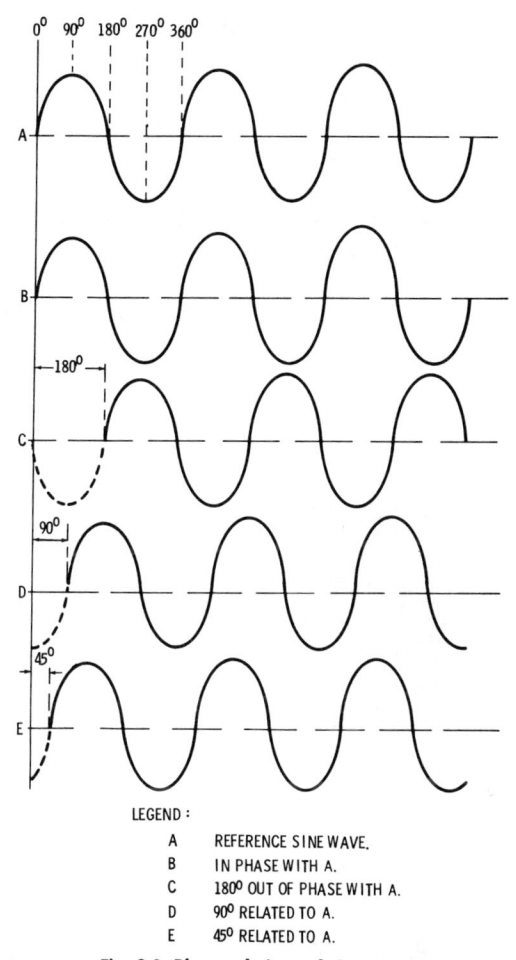

Fig. 2-8. Phase relations of sine waves.

LEGEND:
A REFERENCE SINE WAVE.
B IN PHASE WITH A.
C 180° OUT OF PHASE WITH A.
D 90° RELATED TO A.
E 45° RELATED TO A.

field-effect transistor is a linear device and is useful in the design of low-distortion circuits. Operating characteristics are quite similar to a vacuum tube except for lower supply voltage and no filament.

The understanding of FET operation is aided by a review of the semiconductor-junction theory. You have learned that when a junction is reverse biased, there is no significant movement of charges across the junction. However, as shown in Fig. 2-4B, there is a rearrangement of the electron and hole carriers. The excess electrons of the n segment are drawn away from the junction toward the positive terminal. Likewise, the excess holes in the p segment are drawn away from the junction toward the negative terminal. Hence a major portion of the carriers have been pulled away from the area close to the junction. This is called the *depletion area* (Fig. 2-9). As shown in Fig. 2-9B, a higher reverse bias pulls a greater number of charges away from the junction, and the depletion area widens.

The width of the depletion area in both the p and n segments also depends on the number of carriers within the area, which is a function of the chemical make-up, or *doping,* of the semiconductor material. Since a greater number of excess charges

(A) Reverse biasing.

(B) Higher reverse biasing.

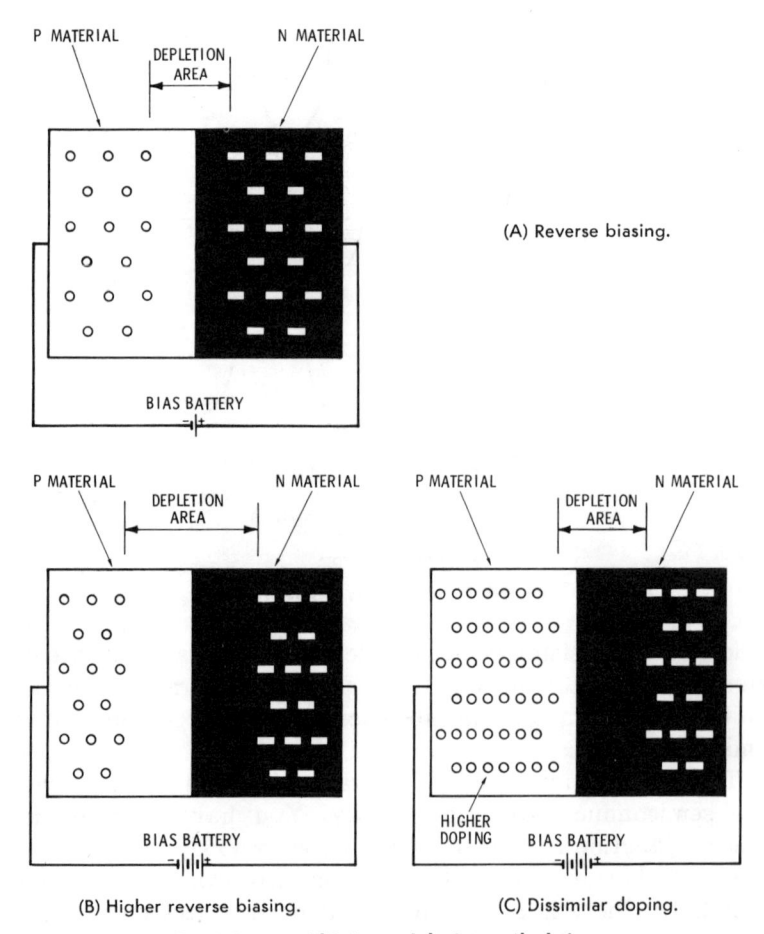

(C) Dissimilar doping.

Fig. 2-9. Influence of biasing and doping on depletion area.

50

are available per volume, the actual depletion region is smaller for a heavily doped material than for one with fewer excess charges (Fig. 2-9C).

The operation of a field-effect transistor depends on the influence that a reverse bias has on the width of a depletion area. A field-effect transistor, as shown in the simple arrangement of Fig. 2-10, is a single-junction (*unipolar*) device consisting of a semiconductor bar (n-type material in the example) into which two facing strips of semiconductor material of the opposite type are diffused (p-type material in the example).

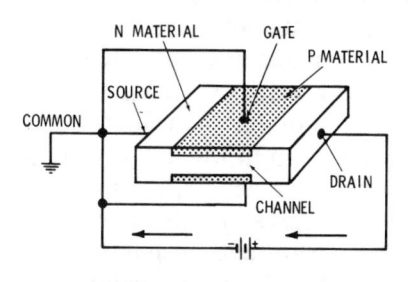

Fig. 2-10. Basic FET construction and operation.

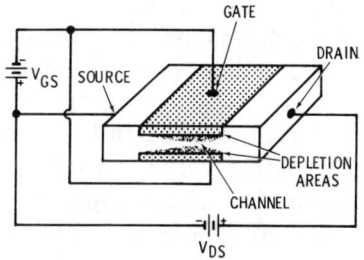

Fig. 2-11. Reverse biasing the FET.

In FET terminology, the bar is called a *channel* while the element that controls the motion of charges along the channel is known as the *gate*. The common end of the channel is known as the *source*, while the opposite end is called the *drain*. The gate can be loosely compared to the control grid of a vacuum tube while the source and drain approximate the operational positions of the cathode and plate respectively.

When the semiconductor bar is biased as in Fig. 2-10, there is a motion of electrons from source to drain through the bar or channel. In the operation of the field-effect transistor, the gate is able to control this motion of charges along the channel. Let us consider how this is accomplished.

The gate-to-channel junction is reverse biased with the source being used as the common connection (Fig. 2-11). This reverse biasing of the junction causes a depletion region to extend into the channel. The presence of the depletion region in the channel causes an increase in the resistivity of the bar (decrease in conductivity). By increasing the gate bias there is a further rise in the channel resistance and the motion of charges (drain current) is reduced. If the gate bias voltage is made high enough, the channel or drain current can be reduced to zero (cutoff).

FET AC OPERATION

An applied audio or other signal causes the gate voltage to vary about the gate bias set by the gate battery voltage. This gate current variation will cause a similar variation in the channel or drain current. In fact, as the gate voltage is made to vary with signal there results a substantial change in the drain current. The actual output voltage results from the variations of the drain current in the drain load resistor (Fig. 2-12). The actual voltage gain of the stage is the ratio of V_o/V_{in}.

A positive swing of the gate voltage produces an increase in the drain current. In turn, there is a drop in the drain voltage (the output voltage swings negative). Conversely, the negative alternation of the input voltage decreases the drain current and there is a positive swing of the drain and output voltage. Input and output voltage are said to be of opposite polarity or *out of phase*.

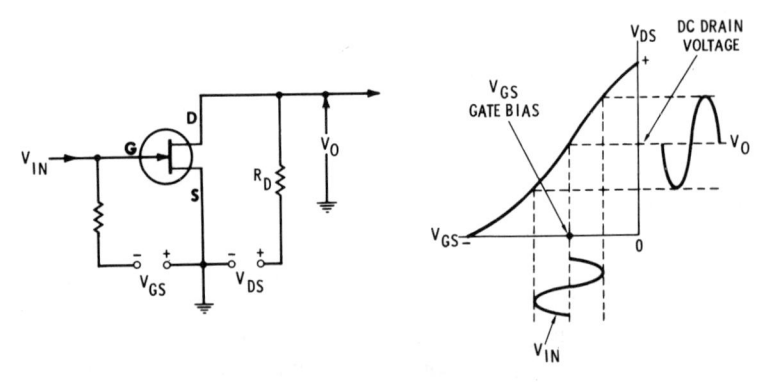

Fig. 2-12. Basic operation of FET amplifier.

DEMONSTRATION 1
Rectifier Operation

In the demonstrations of Chapter 2, use the basic ac-dc pegboard of Chapter 1. The required components are given in the parts list of Chapter 1.

In Demonstration 1, rectifier principles are demonstrated. A complete low-voltage power supply is added in Demonstration 2. This power supply will then be used to supply voltage to a basic transistor circuit in Demonstration 3.

Introduction

A rectifier permits current flow only in a single direction. Its resistance depends on the polarity of the applied voltage. When the polarity of the voltage is such that a positive difference of potential is present between anode and cathode, the resistance is low, and a strong current flows. If the voltage between anode and cathode is negative, the rectifier displays a high resistance, and, therefore, little or no current flows through the rectifier.

A rectifier is useful in converting ac electricity into dc electricity. Most of the electronic equipment in your house employs rectifiers to convert the incoming 110-volt ac power to dc power. In this demonstration, you will learn some of the characteristics of a single silicon rectifier.

Procedure

1. Connect the circuit of Fig. 2-13 across clips 3 and 4 of the pegboard. Connect a jumper between clips 3 and 5. Connect the 68-ohm, ½-watt resistor between clips 4 and 8. Connect the silicon rectifier between clips 7 and 8, with the anode side connected to clip 8. Connect a jumper between clips 6 and 7.
2. Insert the vom between clips 5 and 6. Set the vom to measure current, using a current scale of 0-100 milliamperes (mA) or 0-120 mA. Close the knife switch to the dc side. Record the current flow that is indicated by the meter.
3. Reverse the silicon rectifier, connecting its cathode to clip 8. Close the knife switch to the dc side, and measure the current flow. Notice that it is necessary to set the vom to its most sensitive scale in order to obtain just a slight indication of reverse current flow. It is apparent therefore, that the rectifier permits current flow in only one direction.
4. Reconnect the silicon rectifier with its anode connected to clip 8. Place a jumper between clips 5 and 6 and remove the vom from between these clips. Close the knife switch to the dc side. Measure the voltage drop across the resistor and

53

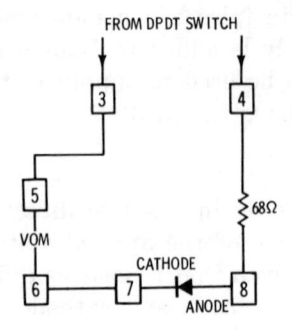

Fig. 2-13. Simple rectifier circuit.

across the rectifier. What is the sum of the two voltage drops?

5. Reverse the silicon rectifier so that its cathode is connected to clip 8. Measure the voltage drops across the resistor and the silicon rectifier.

In the last two steps, the resistor was used as the output of the rectifier circuit. Output existed only when the silicon rectifier was forward biased.

6. Reconnect the silicon rectifier with proper polarity. Substitute a 1000-ohm resistor for the 68-ohm resistor between clips 4 and 8. Close the knife switch to the dc side. Measure the voltage drop across the resistor and the rectifier. Note that the two voltages measure just about the same as they did in Step 4. This proves that despite the very much lower current drawn, the output voltage remains about the same. This indicates that the *regulation* is good.

Use Ohm's law to calculate the dc current flow in the circuit. Current flow is the same in all parts of a series circuit. Thus you need only determine how much current flows through the resistor:

$$I_R = \frac{E_R}{R}$$

where,

I_R is the current through the resistor in amperes,
E_R is the voltage across the resistor in volts,
R is the resistance of the resistor in ohms.

Prove the calculation by removing the jumper from between clips 5 and 6 and measuring the dc current flow with the vom.

7. *For this check, no voltage is applied to the circuit.* Set the dpdt switch to the center position. Set the vom to an ohmmeter range. Use a scale that can read up to approximately 1000 or 2000 ohms. Place the ohmmeter test leads between clips 7 and 8.
8. If the meter reads less than 1000 ohms, it is measuring the forward resistance of the rectifier. If the meter swings up to maximum, it is reading the very high reverse resistance of the rectifier. You can go from one reading to the other by reversing the test lead connections between clips 7 and 8.
9. A rectifier should have a low resistance in one direction and a very high resistance in the reverse direction. If a rectifier should happen to have a low resistance value in both directions or a high value of resistance in both directions, this is an indication that the rectifier is faulty.
10. Reconnect the 68-ohm resistor between clips 4 and 8. Connect the vom test leads between clips 5 and 6. Connect the silicon rectifier with its anode at clip 8. Set the vom to measure dc current.
11. Throw the knife switch to the ac position, and measure the current flow.
12. Reverse the silicon rectifier connections between clips 7 and 8. Throw the knife switch to the ac position. What happens to the deflection of the current meter? Reverse the vom leads connected between clips 5 and 6.

    This step proves that regardless of the manner in which the rectifier is connected, there will be a current flow with the application of ac. This is because the applied ac voltage swings both positive and negative.

    However, the polarity of rectifier connection does influence the direction of dc current flow. When the rectifier was connected with its anode to clip 8, the current flow through the rectifier was from clip 7 to clip 8. With the diode connected in the opposite manner (anode to clip 7), the current flow was from clip 8 to clip 7. Thus the current flow through the meter was also in the opposite direction, and, to obtain a reading, it was necessary to reverse the meter leads.

13. Place a jumper between clips 5 and 6. Connect the silicon rectifier so that its anode is attached to clip 8. Measure the

dc voltage drop across the resistor. Notice that the positive side of the vom (red lead) must be connected to clip 4 in order to obtain a voltage reading.
14. Reverse the silicon rectifier connection between clips 7 and 8. Close the knife switch to the ac side, and measure the dc voltage across the resistor. Notice again that the meter deflects in the improper direction, and it is therefore necessary to reverse the meter leads to measure the voltage drop across the resistor. This time the red lead must be connected to clip 8.

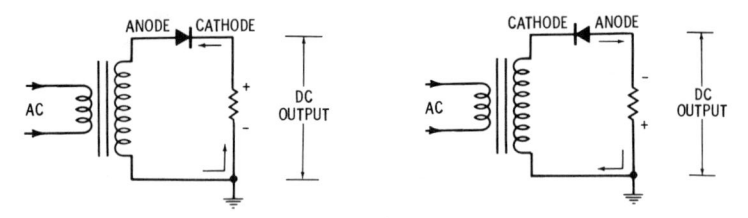

(A) Polarity with cathode toward load. (B) Polarity with anode toward load.

Fig. 2-14. Determination of polarity of dc output voltage.

The preceding two steps demonstrate that a rectifier circuit can be arranged to supply either a positive voltage or a negative voltage with respect to ground (Fig. 2-14). Current flow in either case is in only one direction. In Fig. 2-14A, the current flow is such that a positive dc voltage is present at the top of the resistor. In Fig. 2-14B, the current flow is in the opposite direction, and a negative dc voltage is present at the top of the resistor.

DEMONSTRATION 2
AC-to-DC Power Supply

Rectification is only one step in the conversion of ac to dc power. A transformer is usually associated with the power supply. Its function is to step up or step down the voltage of the ac power source to a level that will produce a desired dc output voltage. Fuses and switches are usually associated with power supplies and, at times, with other controls that may be used to regulate the magnitude of the ac input voltage and/or dc output voltage.

Introduction

Although a rectifier changes ac current to dc current flow, the dc current is not constant. It varies throughout the alternation of the applied sine-wave voltage. This is called a *pulsating* dc output. *Filters* are used to smooth out this variation to obtain a constant dc output. For low-current and low-power applications, filters composed of capacitors and resistors can be employed. For high-powered operation, filter chokes are used instead of resistors. A choke, because of its high inductance, is a good filter, offering high opposition to any change in current.

In this demonstration, a half-wave rectifier and resistor-capacitor filter circuit is connected to the 6.3-volt secondary of the filament transformer on the project board. Its output will be used as a source of dc power for operating several transistor devices to be constructed in succeeding chapters. It will take the place of a battery.

In this demonstration the pegboard will be rearranged considerably. The double-pole, double-throw switch is removed, and the power supply is constructed on the right-hand side of the board. At the conclusion of Demonstration 2, the power supply will be wired on a permanent basis, and the dc voltage made available will be used to power succeeding projects. The basic arrangement is shown in Fig. 2-15.

Procedure

1. Connect one side of the transformer secondary to clip 1 and the other side through the fuse to clip 2. Connect the silicon rectifier between clips 1 and 3 with the anode connected to clip 1. Connect a 220-ohm, 2-watt resistor between clips 3 and 4.
2. Insert the vom between clips 2 and 4 to measure the rectifier-circuit current. Turn on the supply and measure the dc current flow. Record this value.
3. Place a 100-μF capacitor between clips 3 and 4. Connect the plus terminal of the capacitor to clip 3. Note the increase in the dc current flow. The capacitor filters out the pulsating dc and makes a more constant dc voltage available across the resistor, hence the increase in dc current flow. Record the value of current just observed.

4. Place a wire jumper between clips 2 and 4. Measure the voltage drop between clips 3 and 4, and record the value. Remove the capacitor, and take the same reading. The capacitor charges up to the peak value of the ac voltage; hence, a higher dc voltage is made available with the capacitor in the circuit. The capacitor and the resistor together operate as a simple filter.
5. The ac voltage can be measured by inserting the 0.1-μF capacitor between clip 3 and the red test lead of the vom, as shown in Fig. 2-16. Measure the ac voltage with the 100-μF capacitor out of the circuit and with it in the circuit. Record the values. Notice the substantial reduction in the ac-voltage component with the addition of the capacitor, proving its filtering action.
6. Wire a filter circuit to the output of the rectifier. This is done by placing the 220-ohm resistor between clips 3 and 5. Connect one of the 100-μF capacitors between clips 3 and 4; connect the second 100-μF capacitor between clips 5 and 6. Connect these capacitors with their plus terminals to clips 3 and 5. Connect a 1000-ohm, $\frac{1}{2}$-watt load resistor between clips 5 and 6. Connect clips 2, 4, and 6 together with jumpers.
7. Measure the dc voltage at the input (between clips 3 and 4) and output (between clips 5 and 6) of the filter. Notice that with the filter present there is some drop in the voltage that is available at the output of the filter. This results

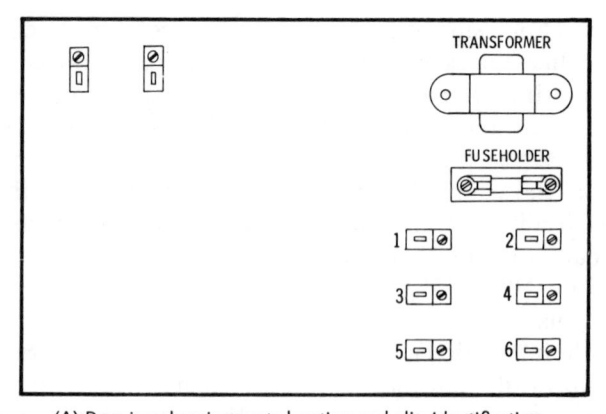

(A) Drawing showing parts location and clip identification.

Fig. 2-15. Rectifier-

from the current flow through, and voltage drop across, the filter resistor.
8. Measure the ac voltage (using the 0.1-μF series capacitor) at the input and the output of the filter. Notice that the ac voltage is very low and barely causes a perceptible reading at the output of the filter. Thus, almost a pure dc voltage is made available between clips 5 and 6.

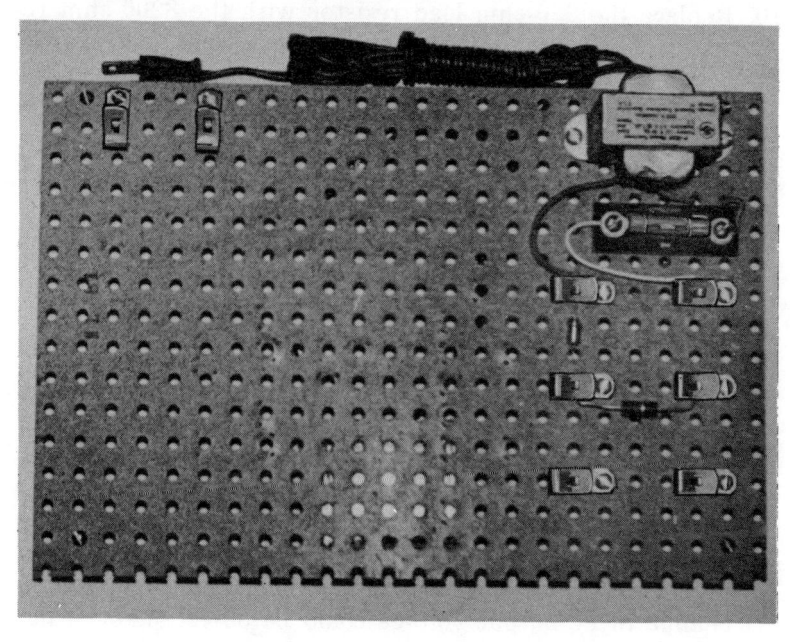

(B) Photograph of pegboard arrangement.

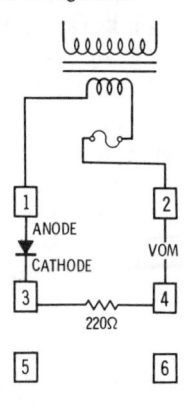

(C) Circuit diagram.

filter demonstration.

9. Remove the 1000-ohm load resistor and replace it with a second 220-ohm resistor. Measure the dc and ac output voltage. The presence of the lower-value resistor is the same as placing a heavier load on the output of the power supply. Thus more current is drawn, and the dc voltage made available has a lower value. The ac component or *ripple* tends to increase with a heavier load on the supply.
10. Replace the 220-ohm load resistor with the 8200-ohm resistor. This resistor represents a very light load on the power supply. Measure the dc and ac voltage across the output. The dc voltage is quite high and the ac voltage very small because of improved filtering action.

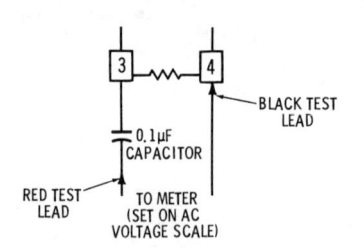

Fig. 2-16. Method of measuring ac when dc is also present.

In summary, the last steps demonstrate that the dc voltage made available at the power-supply output varies with the load placed on the supply, being lower for a heavier load. If the load is reasonably constant, the so-called poor regulation of the supply is not too important. However, the change in dc output voltage could be objectionable when the device being supplied with power displays a changing load. More elaborate power supplies and filters are capable of supplying dc voltages that change very little with changes in load. This type of supply is not required for the low-power transistors used in the projects in this book.

11. Wire a permanent setup to supply a negative dc output voltage. The negative output voltage is more suitable for use with the pnp transistors that will be used in this and succeeding projects. A schematic diagram of the final power-supply wiring is shown in Fig. 2-17. Be sure to connect the rectifier and the capacitors with the proper polarity.

After the supply has been put in its final form, check the ac and dc output voltages. The ac component should be so

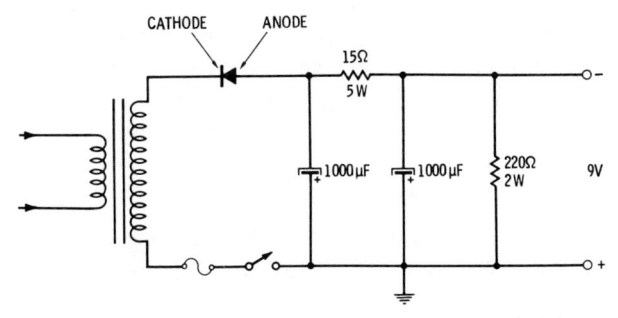

Fig. 2-17. Schematic diagram of transistor-circuit power supply.

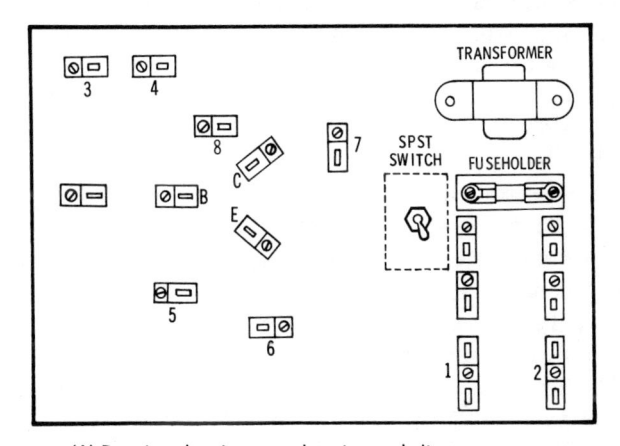

(A) Drawing showing parts location and clip arrangement.

(B) Circuit diagram of power supply.

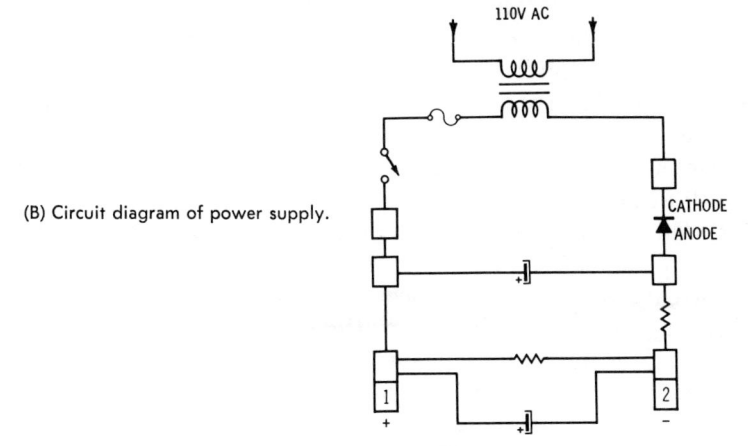

Fig. 2-18. Pegboard dc power supply.

slight that there will be only a barely perceptible reading on the most sensitive ac scale of the vom. The dc output voltage should be approximately 8 to 9 volts negative. The final arrangement of the power supply plus a grouping of clips that will be used in Demonstration 3 is shown in Fig. 2-18.

DEMONSTRATION 3
Bipolar Transistor Operation

The bipolar transistor has revolutionized many phases of the electronics industry. It is efficient, small, and lightweight, and, in most applications, places only a small load on its source of power. It contains two rectifying semiconductor junctions with one of its elements, the base, common to both junctions.

Introduction

If the collector-base junction is reverse biased (Fig. 2-6), there will be no collector-current flow when there is no bias present across the emitter-base junction. Likewise if the base-emitter junction is reverse biased, there will be no collector- or base-current flow. Only when the base of the emitter junction is forward biased is there both base- and collector-current flow. The amount of collector-current flow depends on how much the base-emitter junction is forward biased.

The transistor is basically a current amplifier. When the bias on the base-emitter junction is increased, there is a higher base-current flow. In turn, an increase in the base current produces an amplified current flow in the collector circuit. This is a result of the mass movement of charges across the base-emitter junction when it is forward biased. These charges are attracted to the collector because it is at such a potential that it will attract them as they move across the base-emitter border. This very great attraction of the collector results in a higher collector-current flow than base-current flow. The collector-current flow, in fact, becomes an amplified version of any small-signal, base-current variation.

The transistor is an efficient and effective electronic device. However it is very critical with respect to operating conditions. It can be easily damaged by improper operation and careless handling. One must be particularly careful in soldering transistor leads. In the initial projects, connections to the transistor are made using Fahnestock clips. Soldering is not required. However, if you wish to build projects that are soldered, be certain to refer to the soldering instructions that are given in the next chapter.

When high-power transistors are used, they are often mounted on so-called *heat sinks*; these sinks conduct the heat away from the transistor so that it is not damaged. Other heat sinks snap in position over the case of a transistor. Transistors must be operated with proper bias. In particular, you should not forward bias the collector-base junction or exceed the recommended operating voltages for the transistor. A transistor can be made useless in just an instant by applying an improper voltage.

Be certain you know the element leads of any transistor you use. In the transistors used throughout the projects in this book, the leads can be identified by their position and spacing (Fig. 2-19). Remember this when connecting the transistor leads in the pegboard circuits.

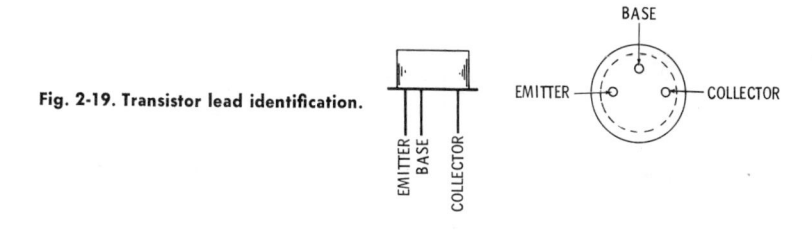

Fig. 2-19. Transistor lead identification.

Procedure

1. Transistor operation will be demonstrated by using an inexpensive general-purpose transistor. Notice in Fig. 2-20 that 3 clips are mounted in a triangular manner at the center of the pegboard. These serve as connections for the base, collector, and emitter leads of the transistor as well as for making external circuit connections.

   Clips 1 and 2 are the source of the + and − collector voltages, and clips 3 and 4 are the source of the + and − base bias. The base bias is controlled with the 25 kΩ potentiometer. Connect the center terminal of the potentiometer to clip 8. Connect the left and right outside terminals of the potentiometer to clips 3 and 4 respectively.

(A) Photograph of pegboard assembly.

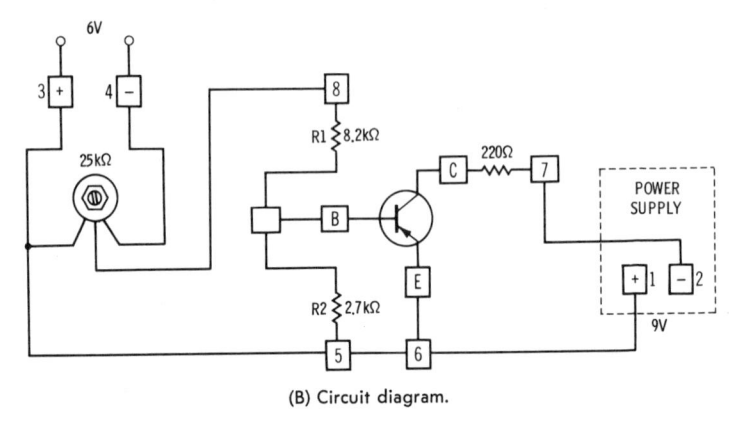

(B) Circuit diagram.

Fig. 2-20. Layout for transistor demonstration.

Connect the 220-ohm collector load resistor between clip 7 and the collector clip. Connect jumpers between clips 5 and 6, and between clip 6 and the emitter clip. Connect clip 1 to clip 6, and clip 2 to clip 7. Connect clip 3 to clip 5. Connect the 8200-ohm resistor (R1) between clip 8 and the base clip. Connect the 2700-ohm resistor (R2) between the base clip and clip 5. Attach the clip-on heat sink to the transistor case.

The base-emitter voltage is obtained from the potentiometer. Resistors R1 and R2 operate as a voltage divider, as shown in Fig. 2-21. The bias voltage applied to this divider determines the base bias current of the transistor. The actual base-emitter bias is much less than the bias voltage applied to the top of resistor R1.

Inasmuch as resistor R1 is a higher value than resistor R2, most of the battery voltage appears across resistor R1. Only a small, fractional part of the battery voltage appears across resistor R2 and on the base of the transistor. In this project, test voltages (at clip 8) of −1, −2, −3, −4, −5, and −6 V will be used.

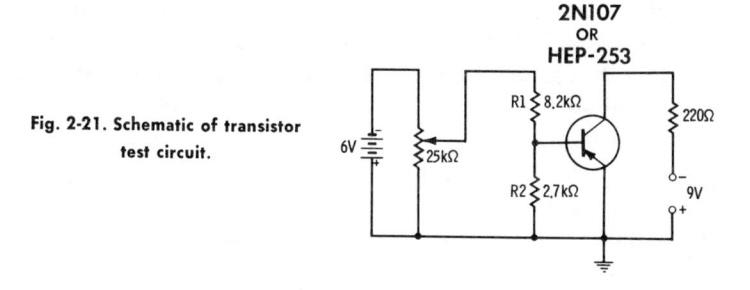

Fig. 2-21. Schematic of transistor test circuit.

2. Adjust the potentiometer for zero bias voltage at clip 8. Turn on the collector power supply. Connect the vom between the collector clip and clip 7. Observe that there is no collector current. Connect the bias battery with opposite polarity: + terminal to clip 4; − terminal to clip 3. Adjust the potentiometer to apply +3 volts to clip 8. Observe that there is no collector current when the base-emitter junction is reverse biased.
3. Connect the battery with proper polarity: + terminal to clip 3; − terminal to clip 4. Adjust the potentiometer to apply −3 volts to clip 8. Measure the collector current by connecting the vom between clip 7 and the collector clip. In this position, the vom shunts out the collector load resistor. Thus, the vom measures the actual collector current of the transistor with the collector-emitter circuit shorted by the meter. Record this meter reading.

Remove the jumper from between clips 5 and 6. Insert the vom in its place and read the base current with a short connected between the collector and clip 7. Record this reading.

4. Calculate the dc beta of the transistor using the formula:

$$h_{FE} = \frac{I_C}{I_B}$$

where,
 h_{FE} is the dc beta,
 I_C is the collector current in milliamperes,
 I_B is the base current in milliamperes.

The value should be between 100-160.
5. Remove the heat sink and, with the vom connected between the collector and clip 7, observe the collector current. Note that the current will start to increase slowly at first, but after a period of time it will begin to increase very rapidly. When this happens, turn the power supply off and replace the heat sink. This effect is known as *thermal runaway* and can burn out the transistor. It proves the necessity for keeping the transistor junctions reasonably cool.
6. Measure the base current and collector current for all bias voltages between 0 and −6 volts. Tabulate the results in Table 2-1.
7. You now have a set of readings that show you how the base current and collector current vary with the applied base voltage. Notice that as the base voltage is made more negative (higher forward bias of the base-emitter junction) the base current and the collector current both increase.

If you now divide a collector current change by a corresponding base current change, you can calculate the ac or small-signal beta of the transistor. Subtract the collector current measured at −1 volt bias from the collector current measured at −3 volts bias. Also, subtract the base current

Table 2-1. Form for Current-Gain Data

Bias Voltage (At Clip 8)	I_C	I_B
0		
−1		
−2		
−3		
−4		
−5		
−6		

measured at −1 volt bias from the base current measured at −3 volts bias. Then divide the collector current change by the base current change to obtain the ac beta. It will approximate h_{FE}, the dc beta. The formula for computing the ac beta is:

$$h_{fe} = \frac{\Delta I_c}{\Delta I_b}$$

where,
h_{fe} is the ac beta,
ΔI_c is the change in collector current,
ΔI_b is the change in base current.

The symbol Δ is the Greek letter *delta* and is used to indicate a change in the quantity that follows it.

8. In the following steps you will determine the voltage gain of the transistor-amplifier stage. The output voltage will be measured between the collector clip and the emitter clip. The input voltage will be measured between the base clip and the emitter clip.
9. Take readings for bias voltages between 0 and −6 volts. Record the readings in Table 2-2.
10. From the data observe that the negative collector-emitter voltage decreases (becomes less negative) with an increase in the negative base-emitter voltage (base becomes more negative). In other words, the change in the collector-emitter voltage is out of phase with the change in the base-emitter voltage.

The actual voltage gain of the stage can now be determined by dividing a collector-emitter voltage change by a corresponding base-emitter voltage change. Subtract the

Table 2-2. Form for Voltage-Gain Data

Bias Voltage (At Clip 8)	V_{CE}	V_{BE}
0		
−1		
−2		
−3		
−4		
−5		
−6		

collector-emitter voltage recorded at −3 volts bias from that recorded at −1 volt bias. Also, subtract the base-emitter voltage recorded at −1 volt bias from that recorded at −3 volts bias. Then, divide the collector voltage change by the base voltage change to obtain the voltage gain:

$$A_V = \frac{\Delta V_{CE}}{\Delta V_{BE}}$$

where,
A_V is the voltage gain of the stage,
ΔV_{CE} is the change in collector voltage,
ΔV_{BE} is the change in base voltage.

A value between 50 and 70 should be obtained.

3

Audio Amplifier and Player

An audio amplifier increases the voltage and/or power level of an applied audio signal. In the usual audio amplifier, the weak audio input signal is built up to a power level that will drive a loudspeaker. The source of the audio signal can be the detector output of a receiver, a microphone, phono cartridge, or the head of a tape player. Microphones, phono cartridges, and tape heads are known as sound transducers. A microphone converts sound variations (or acoustical energy) into an electrical signal. A phono cartridge converts the mechanical undulations cut in a record groove into an electrical signal. A tape head converts the magnetic variations impressed on a tape into an electrical signal. The output voltage variations of such transducers are quite weak and must be amplified before they are applied to a reproducer such as a loudspeaker. A loudspeaker is also a transducer but it converts electrical energy back into acoustical energy.

MICROPHONE

Microphones convert changes of sound pressure into an electrical signal. A common type of microphone is the dynamic microphone (Fig. 3-1). It uses the permanent-magnet, moving-coil principle already considered in connection with meter movements. In this case, the voice or music waves are used to drive a *diaphragm*. The diaphragm is attached to a moving coil. The moving coil is in the field of a permanent magnet. The changes of sound pressure on the diaphragm cause a corresponding back-and-forth movement of the coil in the magnetic field. Consequently, a voltage corresponding to the pressure changes at the diaphragm is induced in the coil. The resultant voltage variation is increased by a step-up microphone transformer.

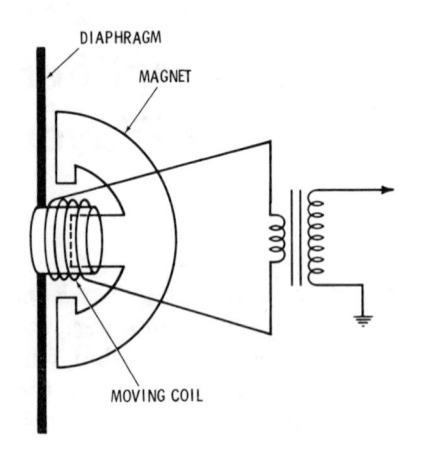

Fig. 3-1. Basic dynamic microphone.

A popular type of microphone uses the piezoelectric principle. In this case a crystal or piece of ceramic that displays piezoelectric characteristics is set into vibration by a diaphragm (Fig. 3-2). When the crystalline material is made to vibrate, it generates a corresponding voltage variation. This variation is amplified as an audio signal.

Still another type of microphone is the carbon unit (Fig. 3-3). A carbon microphone consists of carbon granules packed into a cup or button. A moving diaphragm is attached to one side. As the diaphragm is moved by the sound-pressure changes, the carbon granules alternately compress and separate. As a result, the actual dc resistance of the carbon button varies with the changing sound pressure.

A signal can be formed by passing a dc current through the button. As you know, a change in resistance can cause a change in dc current in an electrical system. Further, if the resistance

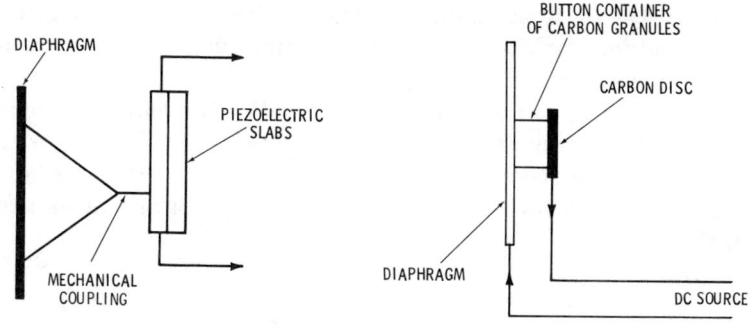

Fig. 3-2. Basic crystal microphone.

Fig. 3-3. Basic carbon microphone.

is varied by the sound-pressure changes, a changing, or ac, current results.

PHONO CARTRIDGE

The basic principles of various types of phono cartridges are very similar to those of microphones. Two common types are shown in Fig. 3-4. Crystal and ceramic cartridges use the piezoelectric characteristic (Fig. 3-4A). In a phono cartridge, a stylus or needle is used instead of a diaphragm. In association with a tone arm and turntable, the stylus is made to follow a record groove as the record rotates beneath the stylus. The record groove has been cut with mechanical undulations that follow the music and voice variations that occurred during the recording session. Thus, as the stylus rides the record groove, it is made to vibrate in accordance with these mechanical undulations.

The stylus conveys these changes as mechanical pressure changes to the crystal or ceramic element. The vibrations of

(A) Crystal or ceramic cartridge.

(B) Moving-coil cartridge.

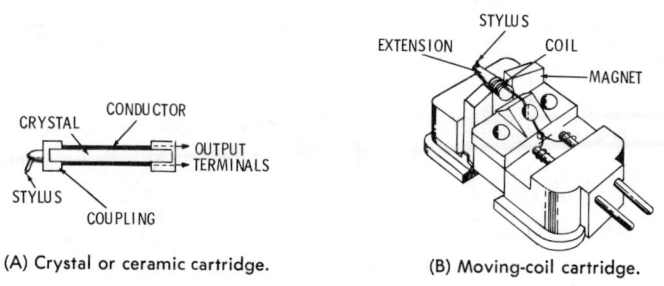

Fig. 3-4. Phono-cartridge principles.

the crystal in turn cause a corresponding voltage variation. This voltage variation can then be amplified and used as an audio-drive signal for a reproducer.

In a moving-coil cartridge (Fig. 3-4B), the coil is attached to an extension of the stylus. As the stylus is vibrated in the grooves of the record, the coil is moved in the magnetic field. As the lines of force are cut by the coil, a voltage is induced in it.

AUDIO AMPLIFIER

A typical audio amplifier is shown in Fig. 3-5. Only a very weak audio signal is available at the input of the first stage. The first stage operates as a voltage amplifier, and the second is a power-output stage.

An amplified change in the collector current is produced when the input signal varies the base current. This current variation produces a similar voltage change across collector load resistor R4. Hence, a substantial voltage variation will be present at the output of the first stage and at the base of the output stage.

As you learned, there are dc and ac collector-current components. In Chapter 1 you also learned that a capacitor blocks the flow of dc current. Thus the dc current component does not appear in the base circuit of the output transistor because of the presence of capacitor C6. However, capacitor C6 has a high capacitance, and it offers very little reactance to the transfer of the audio signal voltages. Almost the entire audio signal voltage appears at the base of the output transistor.

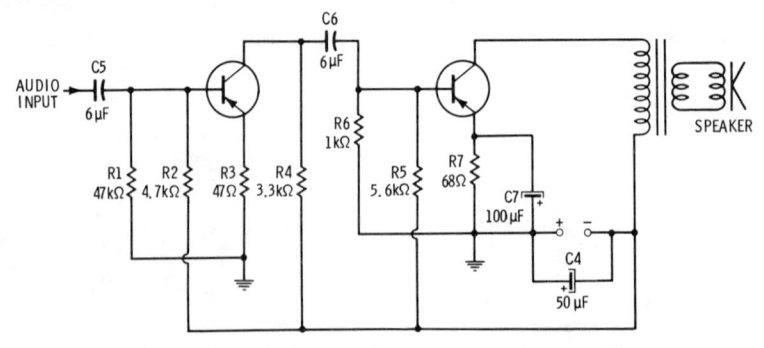

Fig. 3-5. Schematic diagram of two-stage transistor amplifier.

Before considering the output transistor stage, notice the biasing of the input stage. The collector receives its potential from the negative-voltage side of the battery. Since a pnp transistor is being used, the collector is reverse biased. The input emitter junction is forward biased.

The transistor output stage is also biased by the battery. A negative potential is applied to the collector; the positive side of the battery is connected to the emitter circuit. The negative side of the battery is also connected to a voltage divider composed of resistors R5 and R6. The resistor values are chosen so that the base is made negative with respect to the emitter by the proper voltage. In so doing, the base is forward biased (made negative with respect to the emitter) for proper operation.

The negative voltage present at the collector is substantially higher than the negative base voltage. Therefore, the collector is negative with respect to the base, and this junction is reverse biased. It is important to realize that the negative voltage applied to the base circuit of a pnp transistor must not be greater than the negative voltage applied to the collector. If such were the case, the base would actually be negative with respect to the collector, and this junction would be forward biased. This could damage the transistor.

The signal voltage from the input amplifier is applied to the base-emitter circuit of the output stage. The signal-voltage variations cause a like change in the base-emitter current. In other words, the base-emitter current now varies up and down from the dc value of base-emitter bias.

As a result of the base-current variation, there is an amplified collector-current variation. The two stages build up the signal to such a strong level that the current variations in the voice coil cause movement of the voice coil and the cone.

SPEAKER OPERATION

The audio output stage must develop enough audio power to cause a mechanical motion of the speaker assembly. A basic permanent-magnet speaker arrangement is shown in Fig. 3-6. Note that the moving coil, referred to as a *voice coil*, is located in the field of a permanent magnet. Attached to the moving coil is a *cone*. When there is physical motion of the voice coil, the cone is made to vibrate. The vibrating cone, in turn, compresses and expands the surrounding air particles, creating sound waves.

What causes the voice coil to move physically? The magnetic field from the permanent magnet is fixed and unchanging. However, there is also a magnetic field surrounding the voice coil. This field varies with the audio output of the transistor output stage. The two magnetic fields interact and cause a physical movement of the voice coil.

The signal applied to the voice coil is an ac audio variation and, as a result, the coil vibrates back and forth as it follows the positive and negative alternations of the sound signal. The in-and-out movement of the voice coil causes a similar vibration of the speaker cone. This vibration, in turn, sets the surrounding air particles into vibration, and a conversion has been made from an audio-current variation to a sound wave.

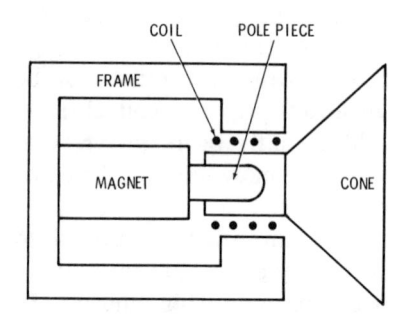

Fig. 3-6. Basic structure of a permanent-magnet speaker.

PUSH-PULL AMPLIFIER

A push-pull transistor stage (Fig. 3-7) is capable of delivering substantially more than twice the audio power output of a single-ended stage using the same type of transistor. A two-fold increase in power output can be anticipated because two transistors are being used instead of one. However, in a transistor push-pull stage it is possible to bias the base-emitter junctions in a manner that will provide additional output. This type of biasing introduces distortion in single-transistor operation.

In the circuit, a voltage divider combination (resistors R1 and R2) is used to set the forward base bias. This bias is lower than would normally be used in a single stage. Low-value emitter resistors (R3 and R4) are often used, particularly when high-power audio transistors are employed. When the operating temperature of a transistor rises above a certain value there is an increase in current flow which further heats the transistor and raises the temperature. The new temperature increase causes a further rise in current, etc. This condition is known as *thermal runaway* and can result in the destruction of the transistor. The use of an emitter resistor causes a bias change and a compensating influence on the current rise. Thus thermal runaway is avoided.

In the usual push-pull transistor stage, an input transformer is used with its secondary center-tapped. By so doing, equal-amplitude but opposite-polarity signals are applied to the separate transistors, as shown in Fig. 3-7. The biasing is such that one transistor conducts for one alternation of the applied sine wave, and the second transistor conducts for the other alternation. The negative alternation of the sine wave applied to the top transistor causes the base-emitter current of this transistor to increase in the same way that the negative base voltage increases. The base-emitter current causes a similar variation in the collector current. Simultaneously, a positive alternation is being applied to the lower transistor, and its collector current decreases. However, the positive alternation is followed by a negative alternation that increases the base-emitter current on the lower transistor. This produces a like change in the collector current of the lower transistor.

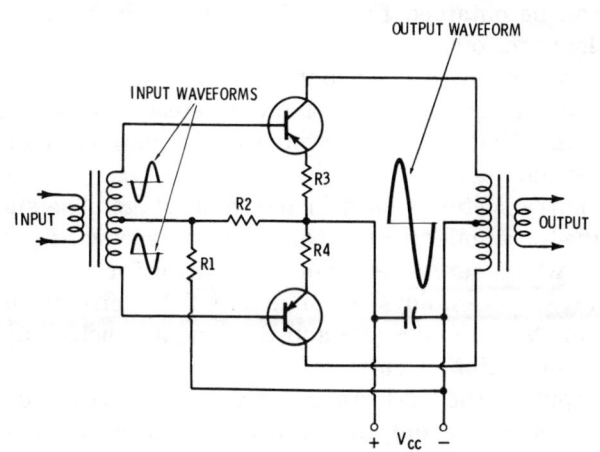

Fig. 3-7. Push-pull transistor audio amplifier.

The output is taken between the collectors and appears across the primary of the output transformer. The top transistor contributes the positive alternation of the output voltage; the bottom transistor contributes the negative alternation. The combination of the two collector currents and the relatively high input impedance of the transformer from collector to collector result in a relatively large audio voltage across the transformer primary. The output transformer provides the necessary current step-up so that the current, in flowing through the voice coil, provides the necessary power for moving the voice coil. In summary, a push-pull connection provides a higher audio output with less distortion than a single-transistor stage.

DEMONSTRATION 1
Two-Stage Audio Amplifier

A two-stage transistor audio amplifier is to be constructed on the pegboard. A photograph of the pegboard assembly and a pictorial wiring diagram are shown in Fig. 3-8. A complete schematic diagram is given in Fig. 3-9. This audio amplifier will also be used in several of the succeeding projects and it can be mounted permanently.

Introduction

The gain of the amplifier is such that it can be driven with signal from a phono cartridge or tape head. Good speaker volume can be obtained from a four- to five-inch permanent-magnet loudspeaker.

The signal source can also be the output of a high-level microphone, or the output of a radio detector. It is indeed a versatile unit that will serve you well in your electronic experimentation.

A single transistor is used in the input stage. The signal to be amplified is applied to the base circuit by way of potentiometer R1, which serves as a volume control for the amplifier. The input stage employs a common-emitter circuit using a two-resistor bias divider (R2 and R3) and a resistor-capacitor emitter stabilization circuit.

The output of the first stage is transformer-coupled to the input of a push-pull output stage using the same transistor type. A similar base biasing arrangement is employed. A push-

(A) Photograph of pegboard assembly.

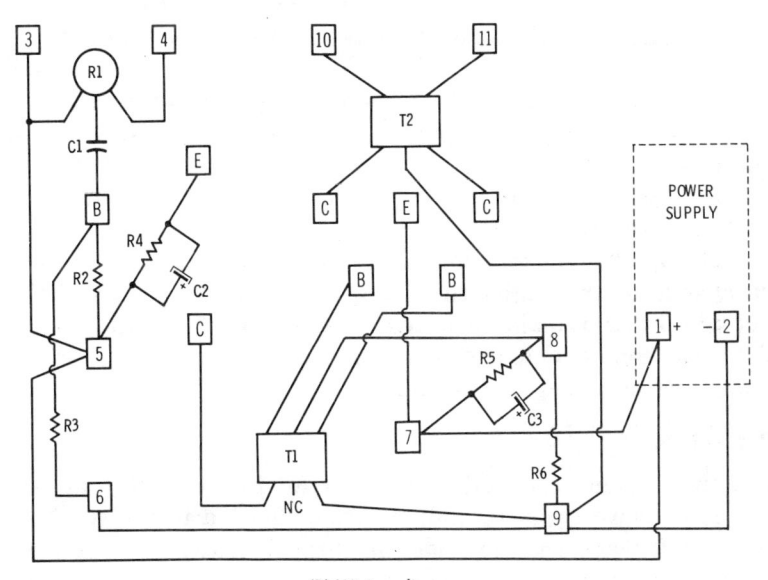

(B) Wiring diagram.

Fig. 3-8. Audio-amplifier demonstration.

pull output transformer matches this stage to a low-impedance loudspeaker. The audio amplifier can be operated from the power supply already wired on the pegboard, or can be operated from an external, 12-volt lantern battery.

77

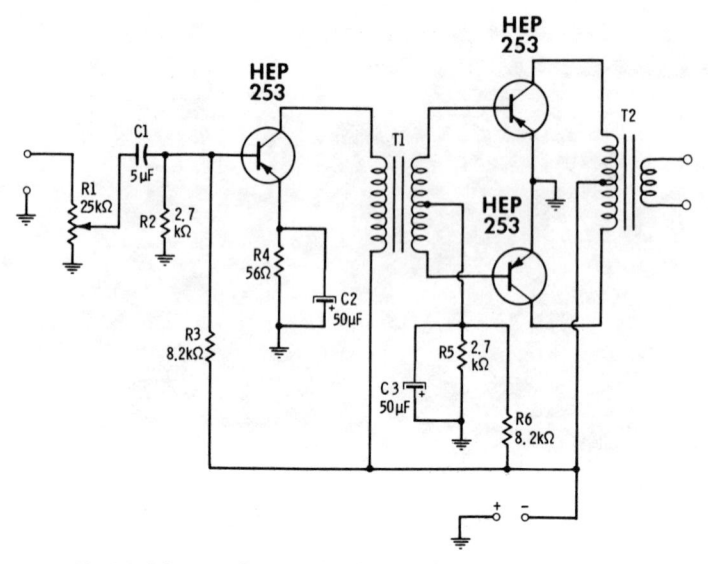

Fig. 3-9. Schematic diagram of audio-amplifier demonstration circuit.

The parts list in Table 3-1 includes the projects for Chapters 3, 4, and 5. Many of the parts used in the demonstrations in the first two chapters can be used again. The pegboard with mounted power supply is used to construct the audio amplifier of this chapter.

Note in Fig. 3-8 that the power supply is left intact on the right side of the pegboard. Likewise, the input clips and the potentiometer remain mounted at the top left of the board. All other components are removed and the construction of the two-stage amplifier can begin.

Procedure

1. First, mount the audio output transformer (T2) at top center. Two clips, 10 and 11, are connected to the two secondary leads that provide the proper match to whatever impedance loudspeaker is employed. If you desire, the loudspeaker can be mounted on an angle bracket and fastened to the pegboard near the output transformer.
2. Mount the five push-pull transistor clips below the output transformer. There are separate clips for the two collectors and the two bases, but a common clip is used for the two emitters. Connect this latter clip to clip 7 and connect clip 7

Table 3-1. Parts List for Chapters 3, 4, and 5

Quantity	Description
1	Pegboard with mounted power supply.
1	Pegboard, 12 in. by 8 in.
30	Fahnestock clips.
1	Lantern battery, 12 V.
1	Output transformer, transistor; primary, 100-ohms center tapped; secondary, 3.2/8/16 ohms (Allied 6-T-52VF or equiv.).
1	Interstage transformer, transistor; primary, 500-ohms center tapped; secondary, 200-ohms center tapped (Allied 6-T-54HF or equiv.).
1	Broadcast antenna coil, universal replacement (Stancor RTC-8736 or equiv.).
1	Crystal diode (1N34A).
3	Transistor (HEP-253 or equiv.).
1	Field-effect transistor (HEP-801 or equiv.).
3	Heat sink, clip-on (TO-5).
1	Permanent-magnet speaker, 5 in.
1	Telegraph key.
1	Potentiometer, 25 kΩ.
1	Variable capacitor, 365 pF.
3	Capacitor, 50 μF, 15-25 V.
2	Capacitor, 5 μF, 15-25 V.
2	Capacitor, 0.05 μF, 15-25 V.
1	Resistor, 1.5 MΩ, ½ W.
1	Resistor, 33 kΩ, ½ W.
3	Resistor, 8200 Ω, ½W.
2	Resistor, 2700 Ω, ½ W.
1	Resistor, 2400 Ω, ½ W.
1	Resistor, 56 Ω, ½ W.

to clip 1 of the power supply. Also, connect clip 3 to clip 5 and clip 5 to clip 1. This completes the common, or ground, system of the amplifier.

3. Connect the two outer leads of the primary of the output transformer (T2) to the collector clips of the push-pull transistors. Connect the center tap of the primary of the output transformer to clip 9, and connect clip 9 to clip 2 of the power supply. Connect one side of the primary of the push-pull input transformer (T1) to clip 9, and then connect clip 6 to clip 9. This completes the collector supply-voltage wiring.

4. Connect the outer leads of the secondary of the push-pull input transformer (T1) to the base clips of the push-pull transistors. Connect the center tap of the secondary of the

input transformer to clip 8. Connect the other side of the primary of the input transformer to the collector clip of the input transistor.

5. In this step, connect the various bias circuits. Connect an 8200-ohm resistor (R6) between clips 8 and 9. Connect a 2700-ohm resistor (R5) between clips 7 and 8. Also, connect one of the 50-μF capacitors (C3) between clips 7 and 8. These three components form the biasing circuit for the push-pull output stage.

Connect an 8200-ohm resistor (R3) between the base clip of the input transistor and clip 6. Connect a 2700-ohm resistor (R2) between the base clip and clip 5. Connect the 56-ohm resistor (R4) between the emitter clip of the input transistor and clip 5. Also, connect a 50-μF capacitor (C2) between the same two clips.

Complete the amplifier wiring by connecting the 5-μF input capacitor (C1) between the arm of the potentiometer (R1) and the base clip of the input transistor.

6. Check your wiring carefully to be certain that it conforms with Figs. 3-8 and 3-9. Set the potentiometer to approximately midposition. Turn on the amplifier. Touch your finger to clip 4 and note that noise can be heard in the loudspeaker. Vary the potentiometer to show how this noise level changes with the setting of the control. The simple technique of touching your finger or a piece of metal to the input of an audio amplifier can be used as a quick check of whether the amplifier is operating, is weak, or is completely dead.

Note that with no input signal, the loudspeaker output is relatively quiet. Momentarily detach the input filter capacitor of the power supply. This results in a substantial 60-Hz hum in the loudspeaker. This is a common audio amplifier fault when there is a bad filter capacitor or other power-supply component, or some undesired 60- or 120-Hz hum pickup somewhere in the amplifier.

DEMONSTRATION 2
Audio Amplifier at Work

Perhaps the most convenient source of signal for the transistor amplifier would be the output of a monaural phono cartridge. Other sources of signal could be a tape head, a microphone, the ear-plug output of a small transistor radio, an audio oscillator, etc. The sensitivity of the amplifier is such that loudspeaker output can be obtained from any one of these sources of signal.

A headset can be used to follow any such signal through the stages of the audio amplifier from input to output. You can also learn how certain defects in the amplifier will affect the quality of the loudspeaker signal.

Procedure

1. Connect the source of signal to the input of the audio amplifier. Connect the grounded side of the signal source to clip 3; connect the ungrounded, or "high," side to clip 4.

    Turn on the signal source and the audio amplifier. Vary the setting of the volume control and observe its influence on the output level.

2. Operate the amplifier and the signal source and note the influence when the input capacitor of the power supply filter is taken out of the circuit. Doing this adds hum to the amplifier output.

3. Momentarily place a short circuit across bias resistor R5. The output level drops and what can be heard is highly distorted. This demonstrates the importance of proper biasing of transistor stages.

4. Disconnect the center-tap lead of the primary of the push-pull output transformer at clip 9. This removes the collector voltage to the output stage and there is no signal output. However, if you now connect the headset between the collector of the input transistor and ground, you can hear the sound.

    The preceding is a common technique for finding faults in electronic circuits. Some sort of indicator such as a meter, oscilloscope, or the simple headset used in this experiment can be used to follow the signal to the point where it drops out. Thus you have a clue as to where the fault lies.

5. Momentarily place a short circuit across resistor R3. This removes biasing from the input stage and also places a shunt on the signal source. Place a short circuit across resistor R4. The sound level drops and is more distorted because of a shift away from the proper operating bias for the input stage.

4

Simple Transistor Radio

Between the sound and the light frequencies is a vast range of frequencies that produce waves that have the ability to travel through space. These are called *radio frequencies*. Although they cannot be seen or heard, radio-frequency waves are all about us carrying voice, music, and other information. With suitable receiving equipment these signals can be detected.

The frequency range between 535 and 1605 kHz (535,000 Hz and 1,605,000 Hz) is the a-m broadcast band. In this frequency range are 107 broadcast channels. Of course, there are more than 107 broadcast stations on the air. Many stations throughout the country operate on the same frequency. Interference between stations on the same frequency is reduced by maintaining distance separation between stations.

When an announcer says, "Tune in again tomorrow on 610," he is referring to the frequency of his particular station—610 kHz. The dial on your radio is calibrated in frequency. As you tune the radio dial, you change the frequency to which the receiver is sensitive; thus, you can tune from one station to another. The receiver is tuned to match the frequency of the station you wish to receive.

AMPLITUDE MODULATION

The radio-frequency signal sent out by a broadcast station is called a *carrier*. The term carrier is used because its purpose is to "carry" audio frequencies (music and voice) between the broadcast station and any receiver tuned to its frequency.

The waveforms of Fig. 4-1 show the use of a radio-frequency carrier. When no voice or music frequencies are being transmitted from the broadcast station, the radio-frequency carrier that is sent out does not change in strength. It is said to have a constant amplitude.

When the announcer speaks into the microphone or a musician plays his instrument, the strength, or amplitude, of the radio-frequency carrier varies up and down. The amplitude changes of the radio wave correspond to the audio-frequency changes of voice and music. The carrier is said to be *amplitude modulated*. In effect the voice and music frequencies "ride" on the radio-frequency carrier.

(A) Audio-frequency signal.

(B) Radio-frequency carrier.

(C) Carrier modulated by audio-frequency signal.

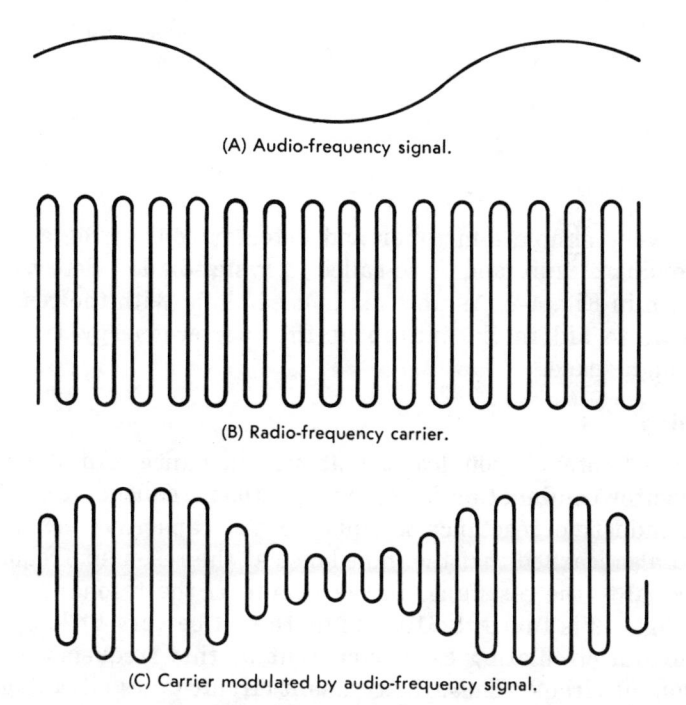

Fig. 4-1. Formulation of an amplitude-modulated carrier.

At the receiver the audio-frequency variations are taken from the carrier. You can then listen to them by means of a speaker or headset. You hear the sounds just as if you were present in the broadcast studio.

RECEIVER PRINCIPLES

The receiver has two major jobs to do. First, it must pick up the desired carrier and keep it separated from others that may be present on nearby frequencies. This selection is made by resonant receiver circuits. The second major task is to remove the low-frequency voice and music variations from the radio-frequency carrier. This process is called *detection* or *demodulation*.

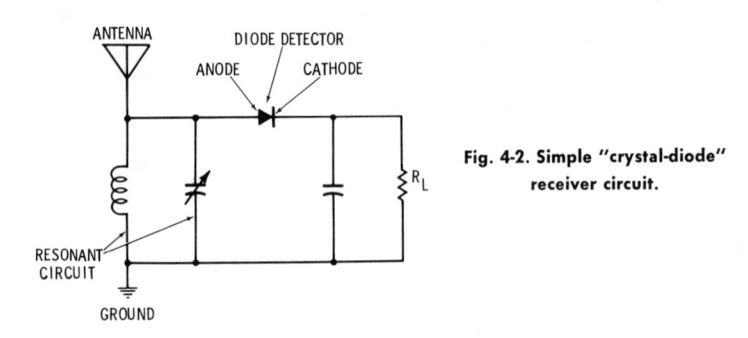

Fig. 4-2. Simple "crystal-diode" receiver circuit.

A very simple a-m tuner and detector consists of a single resonant circuit and a so-called "crystal-diode" detector, as shown in Fig. 4-2. An antenna is used along with the resonant circuit to aid in the pickup of the carrier energy traveling through space.

Tuning

In Chapter 1 you learned about resonance. An inductor-capacitor combination is resonant at that frequency for which the inductive reactance is equal to the capacitive reactance. You also learned that the impedance at the resonant frequency is equal to the resistance in the circuit. If the resonant circuit of Fig. 4-2 is tuned to 610 kHz there will be very little opposition to a circulating carrier current of this frequency in the resonant circuit. Thus, the resonant circuit takes advantage of a 610-kHz signal being picked up by the antenna.

Other broadcast-station carriers are also being picked up by the antenna; however, these carriers are on frequencies other than 610 kHz. At these other frequencies, the inductive and capacitive reactances are not equal. The reactance in the resonant circuit opposes circulating currents on these other carrier frequencies.

How is it possible to tune in a station that has a carrier frequency other than 610 kHz? If you wish to receive a broadcast station that transmits on 1060 kHz, tune the variable capacitor. When the capacitor setting is changed, the capacitive reactance also changes because capacitive reactance varies inversely with capacitance. Hence, the frequency at which the inductive reactance equals the capacitive reactance is no longer 610 kHz.

If you decrease the capacitance of the capacitor by opening the plates, you eventually arrive at a point where the capacitive and inductive reactances are equal for a frequency of 1060 kHz. If there is a 1060-kHz carrier present on the antenna, it will now circulate a strong current in the resonant circuit. Since the circuit is no longer resonant to 610 kHz, the 610-kHz station carrier will be rejected.

The inductance value of the inductor and the minimum and maximum values of the capacitor are selected to make the circuit resonant in the range from 535 to 1605 kHz as the capacitor is tuned from its maximum value to its minimum value. In this manner the circuit can be tuned over the entire a-m broadcast band.

Diodes

The signal voltage that is developed across the resonant circuit is next applied to a crystal diode. This diode is part of the detector circuit. It recovers the modulation that has been riding on the radio-frequency carrier.

A diode has two elements, or electrodes—an *anode* and a *cathode*. The anode is on that side of the diode toward which the electrons will flow freely. For this reason it is said to be the positive side of the diode. The cathode is the negative side. When a positive voltage is applied to the anode, making it positive with respect to the cathode, there will be a good flow of current through the diode. If the voltage applied to the diode is such that the anode is made negative with respect to the cathode, there will be no significant current flow.

Detection

The first step in recovering the voice and music frequencies from the modulated radio-frequency carrier is to pass only one side of the modulated rf signal. In the circuit of Fig. 4-2 the positive side of the modulated wave causes diode conduction. Hence, a current flows in the output resistor (R_L).

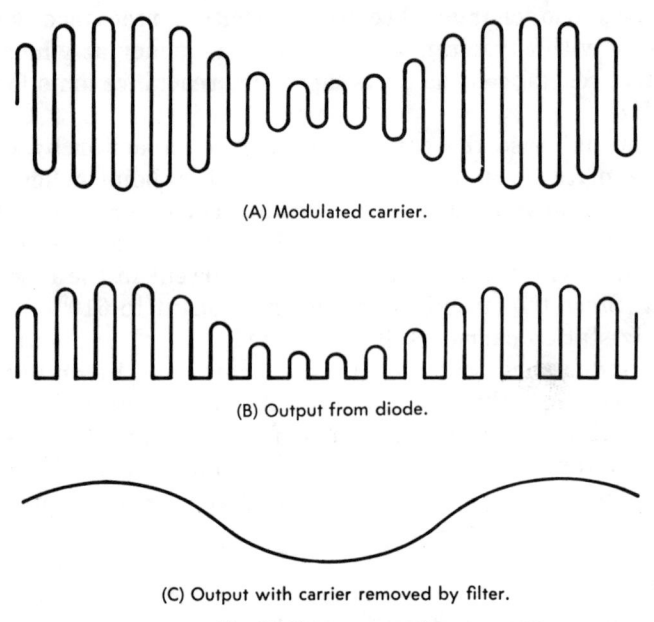

(A) Modulated carrier.

(B) Output from diode.

(C) Output with carrier removed by filter.

Fig. 4-3. Detector operation.

Oppositely, the negative swing of the carrier places a negative potential on the anode, and the diode does not conduct. Thus, the variations in diode current flow will follow the positive variations of the modulated carrier (Fig. 4-3B). The current variations in the output resistance match the voice and music variations on the carrier and develop a corresponding voltage variation across the output.

Notice that a small capacitor is connected across the output of the diode circuit. This capacitor removes (or filters out) the radio-frequency variations. The voice and music frequencies remain and are a good copy of the audio variations picked up by the studio microphone (Fig. 4-3C).

In practice, the output resistor is replaced with a pair of headphones which have a resistance of several thousand ohms or higher. The variations in the diode current through the headphones cause their diaphragms to vibrate accordingly. These vibrations generate sound waves that can be heard when you wear the headphones.

Antenna Requirements

For best results you must do everything possible to deliver a strong signal to the crystal detector. Only then will the detector be able to produce a strong audio variation through the headphones. In general, the longer the receiving antenna, the greater is the signal that can be delivered to the detector of the crystal radio. A good "ground" also helps to improve the pickup sensitivity of the antenna and crystal radio.

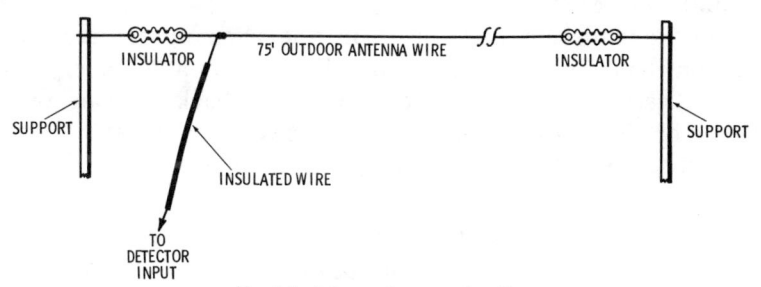

Fig. 4-4. Antenna for crystal radio.

Run a ground wire to a cold-water pipe or a heating-system pipe. If it is convenient, you can run a wire to some outside ground, such as the ground pipe for your television antenna system. Another alternative is to drive a rod 4 to 8 feet into the ground and make your ground connection to the top of the rod.

If you have plenty of space available, the longer the antenna (up to about 100 feet), the better are the results. A 75-foot antenna is a good compromise. The antenna wire is strung between two insulators, as shown in Fig. 4-4. A piece of insulated wire is then run from one end of the antenna down to the input of the crystal radio. Be sure the insulated wire makes a firm metal-to-metal connection with the antenna wire.

A soldered connection is preferable. *Keep your antenna clear of all electric power lines. Do not run it over or under power lines or in a position where it might fall across a power line.*

It is a good idea to install a lightning arrester near the point where the lead-in enters the building. The arrester should be solidly connected to a good ground. If you have had no experience in putting up an antenna, always ask for the guidance and help of someone who has had experience.

An indoor antenna usually performs well if you are not too far from the broadcast station to be received. Even 30 to 50 feet of insulated indoor antenna wire strung around the room will give you some pickup sensitivity for receiving strong local stations.

A better arrangement is to run about 75 feet of indoor antenna around the attic. One end of the wire is continued down into the house and is connected to the detector input. For all indoor antenna installations, make certain you have a good ground connection.

The transmission line for your television or fm receiver can sometimes be used as a makeshift antenna for a crystal radio. Disconnect the twin lead from the back of your television or fm radio, then twist the two ends together and connect them to the input of the crystal radio.

DEMONSTRATION 1
Crystal Radio

In this demonstration, you will construct a crystal-diode detector and a follow-up audio amplifier built around a field-effect transistor (FET). It will be constructed on the second pegboard. The output of this small crystal radio can be applied to the input of the audio amplifier constructed previously. Good loudspeaker volume is obtained. An alternative plan is to use a high-impedance headset at the output of the field-effect transistor.

Introduction

The schematic diagram for the crystal radio is shown in Fig. 4-5. The resonant input transformer is a universal replacement coil. This type of coil is readily available from mail-order houses, and is the type designed for the antenna stage of a-m radios. The tuning capacitor is a standard 365-pF a-m broadcast type.

An advantage of using a field-effect transistor is its very high input impedance. Therefore a high value of gate resistor R1 can be employed. A strong output signal can be derived and a minimum load is placed on the crystal-detector circuit, which improves its sensitivity and selectivity. The field-effect transistor is connected in a common-source circuit, and builds up the demodulated audio signal to about 0.25 volt when receiving a strong signal.

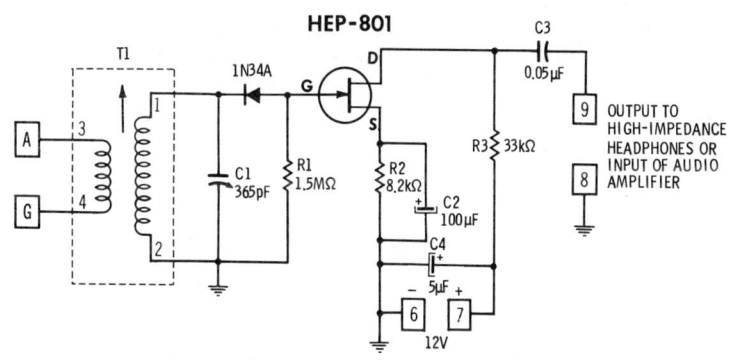

Fig. 4-5. Schematic diagram of small crystal radio.

The performance of this simple crystal radio is very good and a variety of local and long-distance stations can be received, particularly at night.

Procedure

1. A photograph of the pegboard crystal radio and the pictorial wiring diagram are given in Fig. 4-6. As a first step, a set of four leads must be attached to the four terminals of the antenna input coil. The base diagram for the Stancor RTC-8736 is shown. Similar plans are used for other universal replacement coils.

   This transformer can be seen mounted at the upper left-hand side of the pegboard. Connect leads from terminals 3 and 4 of the transformer to the antenna and ground clips. Connect the variable capacitor C1 across the secondary of the transformer with the rotor being connected to ground clip 2 and the stator to clip 3.

(A) Photograph of pegboard assembly.

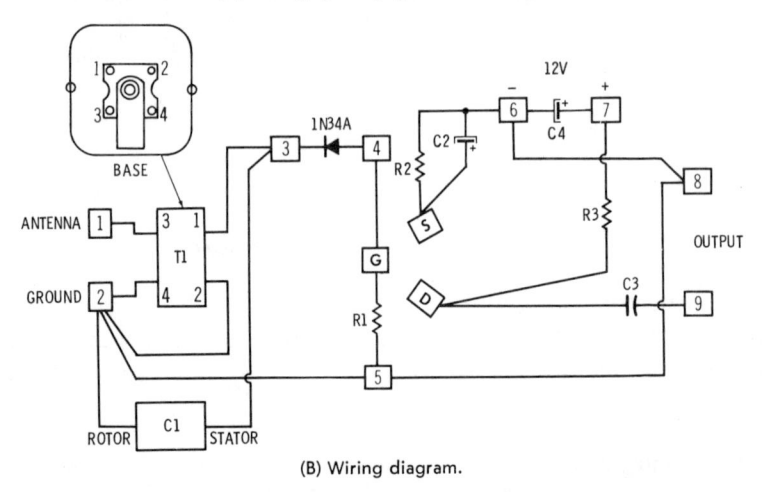

(B) Wiring diagram.

Fig. 4-6. Crystal radio demonstration.

2. Connect the crystal diode between clips 3 and 4 with the anode side of the diode connected to clip 4.
3. The three clips mounted in a triangle are used to mount the field-effect transistor. Clips are provided for gate, source, and drain elements. Small soldering lugs are mounted on the same screw and nut that holds each clip to the pegboard. Solder the leads of the field-effect transistor to these lugs. In making the solder connections, use long-nosed pliers as a heat sink between each transistor

lead and the transistor as the lead end is soldered to the lug.
4. Connect a jumper between clip 4 and clip G (gate). Connect the diode load resistor R1 between the gate clip and clip 5.
5. Connect the 12-volt supply voltage between clips 6 and 7. Connect drain-load resistor R3 between clip 7 and clip D (drain). Connect the source resistor-capacitor combination R2 and C2 between clip 6 and clip S (source).
6. Be certain that the common or ground system is completed. Join together clips 6, 8, 5, and 2.
7. The output of the radio is between clips 8 and 9. Connect capacitor C3 between the drain clip and clip 9.
8. Attach an antenna to the crystal radio. For good results, the antenna should be at least 75-feet long according to instructions given previously in this chapter. Furthermore, a good ground is essential. Run a ground wire from clip 2 to a cold-water pipe, radiator pipe, or an external ground in the form of a pipe or rod driven at least four feet (preferably more) into the ground.
9. Attach a lead between clip 9 of the radio and the ungrounded input (clip 4) of the audio amplifier. Connect a lead from clip 8 of the radio to the ground input clip (clip 3) of the audio amplifier. Turn on the audio amplifier and crystal radio.
10. Vary capacitor C1 over its range. If the antenna is adequate, you will be able to receive quite a number of stations. With the capacitor plates nearly open, you will receive the broadcast stations that transmit at the high-frequency end of the broadcast band between approximately 900 and 1600 kHz. With the capacitor plates almost fully meshed, you will receive the low-frequency stations that transmit from between approximately 540 and 900 kHz.
11. Note that there is a variable adjustment included as a part of the antenna transformer T1. Tune in one of the broadcast stations at the low-frequency end of the band. You should receive this signal with the capacitor plates just about fully meshed.

If this is not so, set capacitor C1 until its plates are almost completly meshed. Adjust the slug in transformer T1 until you hear the lowest-frequency station in your area

come through. In most areas this will be some station broadcasting on 540, 550, 560, 570, 580, or 590 kHz.
12. The highest-frequency signal you will be able to receive will be one on a frequency somewhere between 1400 and 1600 kHz.
13. You can experiment with various antenna lengths and types. It is great fun to stay up late at night and determine the most distant station you can receive.
14. Turn off the audio amplifier and disconnect the leads at clips 8 and 9 of the radio. Connect your headset across clips 8 and 9. You can now receive a number of stations at earphone volume level for quiet listening. The performance you obtain, of course, depends on the sensitivity of your headset. With this arrangement, you can spend your time listening without annoying others in the same room. There is no reason why you can't pick up stations thousands of miles away, particularly late at night.

5

Audio Oscillator

Oscillators are used to generate sine waves and various other types of waves such as square, pulse, and sawtooth waves. Oscillators can be designed for operation on audio or radio frequencies. For example, an oscillator could be used to generate a 60-Hz sine wave or, with proper circuit constants, a sine wave at any frequency in the audio spectrum. Oscillators are also used to generate radio-frequency waves; consequently, they are basic to all types of transmitters. In this chapter you will construct an audio-frequency oscillator. A radio-frequency oscillator will be constructed in the project of Chapter 6.

FEEDBACK

In Chapters 2 and 3 you studied and worked with transistor amplifiers. The signal voltage made available at the collector circuit had a greater amplitude than the signal applied to the base input circuit. In the case of an oscillator, there is no external input signal; the varying signal is formed within the oscillator stage itself. An oscillator is said to be self-generating.

To study the basic principles of an oscillator, let us first assume that there is an input signal. However, instead of being a straight amplifier, the stage has a portion of the output signal fed back to the input (Fig. 5-1). In an oscillator circuit, the feedback arrangement must be such that the feedback voltage reaches the base of the transistor with the same phase as the base signal. By so doing, the signal fed back from the collector circuit reinforces the signal present in the base circuit. Since the input signal now has greater amplitude, there is a further increase in the amplitude of the collector signal. Consequently, the feedback voltage is higher, and there is an additional increase in the base voltage. In fact, the feedback becomes so great that the stage continues to work, or oscillate, even when the external input signal is removed.

Actually it is not necessary that there be any input signal at all. In an oscillator stage there is bound to be some irregularity or change that occurs when the oscillator is turned on. This slight variation is amplified; the feedback circuit plus reamplification soon build up the variation to such a level that self-supporting oscillations exist.

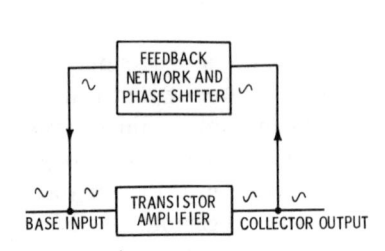

Fig. 5-1. Basic oscillator feedback path.

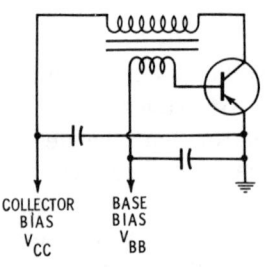

Fig. 5-2. Transformer feedback arrangement.

One might expect that a feedback arrangement would result in a build-up of signal level without limit. However, the transistor is able to handle signals of only a specific amplitude range. Thus the build-up of signal only increases to the level set by the circuit and the operating characteristics of the transistor.

Feedback Polarity

To maintain oscillations, it is necessary that the feedback signal have the same polarity as the base signal. As you learned in Chapter 2, the collector voltage and the base voltage are normally out of phase. Thus, before a transistor stage can oscillate, it is necessary that the feedback arrangement shift the phase of the collector voltage 180°, so that when it arrives back at the base circuit, it will be of the same phase as the base signal voltage.

The basic feedback oscillator of Fig. 5-2 shows how this is accomplished. In this case, the transformer provides the necessary shift in the phase of the feedback voltage. The voltage across the secondary of the transformer is 180° out of phase with respect to the collector voltage variation applied to its primary. Hence, the voltage fed back to the base circuit is in phase with the original change in the base voltage. This is called *positive feedback*.

A few numbers will help you to better understand how oscillations originated. Assume that soon after the oscillator is turned on, a 0.1-volt variation appears at the base. Further assume that the gain of the transistor and the feedback arrangement are such that a 0.1-volt signal is made available across the secondary of the feedback transformer. This 0.1-volt signal is in phase with the original base-voltage change. As a result, the effective base voltage is now 0.2 volt. Since the gain of the amplifier has not changed significantly, and there has been no change in the feedback network, there will now be nearly 0.2 volt present across the secondary. This will add to the previous 0.2 volt and provide almost a 0.4-volt base signal. This base signal is now amplified and fed back to make a higher-level signal across the secondary and at the base. The process continues until the output voltage and the base voltage build up to a limit set by the design of the transistor stage. At this maximum level the stage oscillates continuously, generating an output of a reasonably fixed amplitude. Actually, the build-up of oscillations to a peak value occurs almost instantly when the oscillator is turned on. Likewise, the oscillator output drops to zero very quickly after the source of power is switched off.

You can understand why a 180° phase shift is necessary if you consider what would occur if there were no phase shift introduced in the previous example. In this case, the collector output voltage fed back to the base would be exactly out of phase with the base voltage. If you assume that a 1-millivolt signal appears at the base, there would also be a 1-millivolt feedback signal. Inasmuch as the two voltages have the same amplitude but are opposite in polarity, the net base voltage would fall to zero. Thus there would be no build-up of signal, and the stage would not oscillate. In this case, the feedback component subtracts from the original base-voltage change. This is called *negative feedback*.

Frequency of Operation

As you learned, the frequency of an audio voltage depends on the number of variations that occur each second. In the case of a sine-wave audio oscillator, the frequency of operation is the number of audio sine waves that are generated by the oscillator each second. The oscillations occur at the frequency at which the feedback voltage is exactly in phase with the base voltage. In practice this occurs at only one frequency. At other frequencies, the feedback voltage may be 150°, 30°, or some other value out of phase with the base voltage. Thus there would not be true in-phase feedback. Only at the in-phase frequency do we have maximum feedback and the rapid build-up of oscillations.

It is apparent from the preceding remarks that the actual frequency of operation of an audio oscillator can be set to some specific frequency by making sure that the feedback arrangement shifts the output-voltage phase by exactly 180° at this desired frequency. In the case of the simple oscillator shown in Fig. 5-2, the characteristics of the transformer and, to a lesser extent, the other circuit components determine the frequency at which the feedback is shifted exactly 180°.

AUDIO OSCILLATOR TYPES

Various types of feedback networks can be used for transistor audio oscillators. In Fig. 5-3A, a series of three resistor-capacitor combinations is used to obtain a 180° phase shift of the collector output. Each resistor-capacitor combination results in a phase shift of 60° as the feedback voltage follows the circuit path from collector to base. Therefore, the total change in phase is 180°, and the feedback voltage arrives at the base with the same phase as the base voltage (Fig. 5-3B).

There is only one frequency, however, at which each RC combination provides an exact 60° phase shift. Thus, the values of the resistors and capacitors determine the frequency of operation of the oscillator. If the oscillator is to operate on a different frequency, the values of these components can be

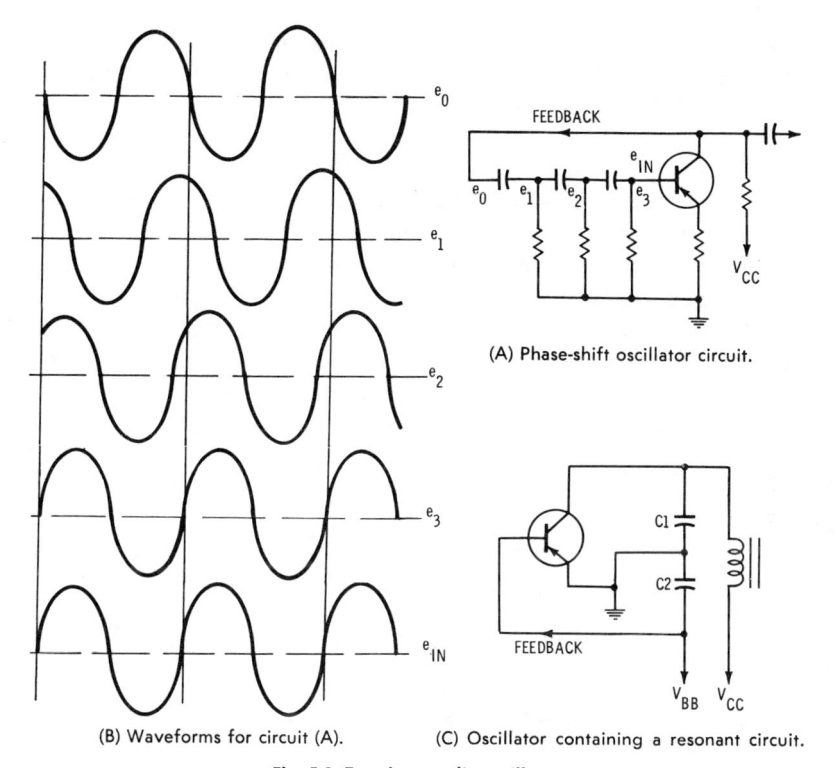

(B) Waveforms for circuit (A). (C) Oscillator containing a resonant circuit.

Fig. 5-3. Transistor audio oscillators.

changed. In fact, if the oscillator frequency is to be made tunable, variable capacitors or resistors can be employed.

Fig. 5-3C shows how an inductor and two capacitors can be used in a feedback arrangement. Actually the combination forms a resonant circuit that determines the audio-oscillator frequency. The portion of the collector-circuit voltage that appears across capacitor C1 is also present between the base and emitter of the transistor. This is the feedback voltage component. Notice that its polarity is such that the base and collector voltages are 180° related, and therefore the feedback voltage and base-emitter voltage are in phase.

The 180° relationship exists only at the frequency of resonance for the inductor-capacitor combination. Hence, feedback is at its strongest, and oscillations are sustained only at this frequency. The frequency of operation can be altered by changing the value of the inductor and/or the capacitor.

There is still another way of obtaining the 180° shift needed for in-phase feedback. Instead of using one transistor stage, it is possible to use two stages as shown in Fig. 5-4. Since each stage provides a phase shift of 180°, the total phase shift is 360°, bringing the output of the second stage in phase with the input of the first stage. This is the relationship that will provide continuous oscillations.

The frequency of operation would be indefinite, and the generated signal would not be a sine wave were it not for the resistor-capacitor network that is inserted between the output of the second stage and the input of the first stage. This network or "bridge" ensures that the oscillator operates only on the frequency determined by the network components.

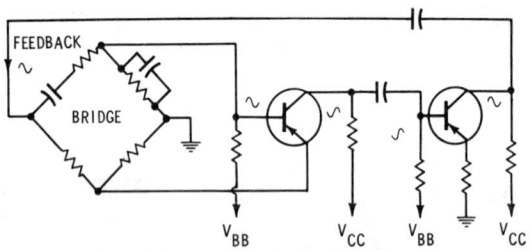

Fig. 5-4. Two-stage feedback oscillator.

DEMONSTRATION 1
Tone Oscillator

A simple tone oscillator can be constructed from the input stage of the audio amplifier constructed in Chapter 3. A few circuit changes and the addition of two components establish the circuit of Fig. 5-5. Note that the supply voltage is now applied to the center tap of the primary of the interstage transformer. The bottom of the primary is connected back to the base through feedback capacitor C4 and resistor R7. For code practice, a telegraph key can be inserted in the same feedback path.

Procedure

1. Disconnect capacitor C1 from the base clip of the input transistor. Disconnect the bottom of the transformer primary from clip 9. Connect the center-tap lead of the primary to clip 9.

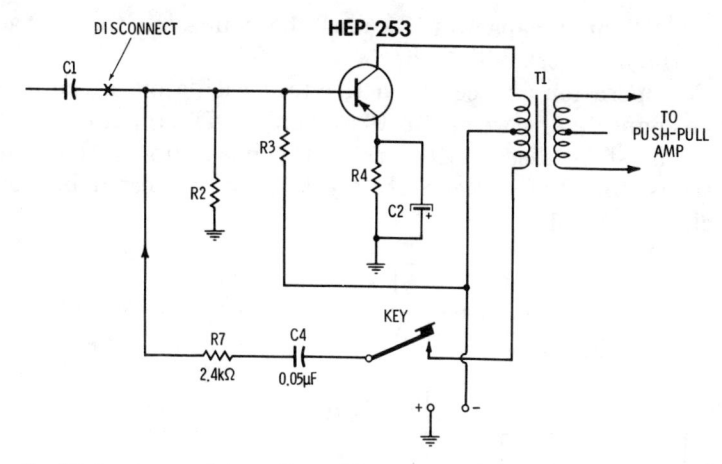

Fig. 5-5. Input stage of the audio amplifier constructed in Chapter 3 (see Fig. 3-9) rearranged to form a tone oscillator.

2. Insert the series feedback resistor-capacitor combination and telegraph key between the bottom of the transformer primary winding and the base clip of the input transistor.
3. Turn on the audio amplifier. You will hear a tone output when the key is held down (feedback circuit closed).
4. Shunt a second 0.05-μF capacitor across the feedback capacitor C4. What happens to the tone frequency? Shunt a 220-ohm resistor across the feedback resistor R7. What happens to the tone frequency? The frequency of the tone is determined by the values used in the resistor-capacitor feedback circuit. The higher the value of the capacitor and the higher the value of the resistor, the lower is the tone frequency. Stated another way, the higher the RC product (time constant), the lower is the operating frequency of the oscillator.
5. If you desire, the tone circuit can be attached permanently to the audio amplifier pegboard with the use of three additional clips, 12, 13, and 14. This arrangement is shown in Fig. 5-6. The network is connected between the base clip of the input transistor and clip 12. The center tap of the transformer primary is connected to clip 13, while the bottom side of the transformer primary is connected to clip 14.
6. When you wish to operate the audio amplifier in a normal manner, a lead must be connected between clip 14 and clip

9. The input capacitor C1 must be connected to the base of the input transistor.
7. For operation as a code-practice tone oscillator, a lead must be connected between clip 13 and clip 9. The input capacitor C1 must be disconnected from the base clip of the input transistor. Your telegraph key is then connected between clips 12 and 14.

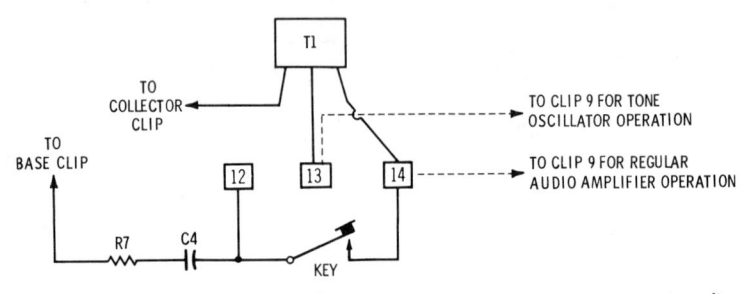

Fig. 5-6. Partial wiring diagram showing connections necessary to convert audio amplifier of Chapter 3 into tone oscillator (refer to Fig. 3-8B).

DEMONSTRATION 2
Code Practice

The earliest method of sending messages by radio used the Morse dot-and-dash code—just as messages were sent over telegraph wires. The radio carrier was turned on and off with a telegraph key to tap out the message. This method of radio transmission is still used today because of its simplicity and ability to penetrate static and other types of interference.

Introduction

Thousands of radio messages are transmitted by Morse code each day. Commercial and pleasure vessels communicate with base and navigation stations by means of code. Coded transmissions are widely used in aircraft communications. Hundreds of thousands of radio amateurs are licensed to transmit code. This fascinating hobby permits radio amateurs to talk to other amateurs at all parts of the globe. Coded transmissions from simple and low-power amateur transmitters travel thousands of miles.

The Morse code alphabet and other important symbols are given in Table 5-1. In sending a *dot*, the telegraph key is pushed

Table 5-1. Morse Code

Letter	Code		Number/Symbol	Code
A	● —		1	● — — — —
B	— ● ● ●		2	● ● — — —
C	— ● — ●		3	● ● ● — —
D	— ● ●		4	● ● ● ● —
E	●		5	● ● ● ● ●
F	● ● — ●		6	— ● ● ● ●
G	— — ●		7	— — ● ● ●
H	● ● ● ●		8	— — — ● ●
I	● ●		9	— — — — ●
J	● — — —		0	— — — — —
K	— ● —			
L	● — ● ●		PERIOD	● — ● — ● —
M	— —			
N	— ●		QUESTION MARK	● ● — — ● ●
O	— — —			
P	● — — ●		COMMA	— — ● ● — —
Q	— — ● —			
R	● — ●			
S	● ● ●			
T	—			
U	● ● —			
V	● ● ● —			
W	● — —			
X	— ● ● —			
Y	— ● — —			
Z	— — ● ●			

down to make a momentary electrical contact and is released immediately. In sending a *dash*, the telegraph key is held down for a length of time equal to three dots. Each time a letter is sent, there is a short space between one letter and the next. After you have sent a word, there is a somewhat longer spacing between the last letter of one word and the first letter of the next. Hence, by regulating the number of dots, dashes, and spacings, you can send coded messages. Messages can be sent at very low speeds, say one word per minute, or up to speeds in excess of fifty words per minute.

To become a radio amateur, it is necessary that you learn to send and receive at a rate of five words per minute. After you reach this speed you will be able to obtain a beginner's, or *novice-class*, amateur license. This permits you to transmit and receive by code on specific frequencies in some of the amateur-radio bands. You then have two years to build up

Fig. 5-7. A typical telegraph key.

Courtesy E. F. Johnson Co.

your code speed to 13 words per minute. This higher rate of speed is required for your *general-class* license.

In using a telegraph key (Fig. 5-7), you should not hold the key firmly. Instead, hold it lightly on the sides with your thumb and your longest finger. The index finger rests lightly on the top of the knob. Place your key far enough in front or to the side of you on the table so that your elbow rests on the table. The dots and dashes are formed by moving your wrist up and down as you press and release the key. Remember that you do not have to pull the key back; it is brought back by a spring when you release your finger pressure.

Learning the code takes a little patience and much practice. It is surprising, however, how quickly you can build up to a code speed of five words per minute.

Procedure

1. Hold your telegraph key properly, press the key down and release it quickly. You have sent a single dot, or the letter "E." Many persons learn the code by studying the letters in related code groups that are easy to memorize. For example, two groups that are easy to memorize are:

    | E . | T – |
    | I . . | M – – |
    | S . . . | O – – – |
    | H . . . . | |

    Send two dots, pause a moment, and then send three dots. In so doing you have transmitted the word "IS." Now send a dash by holding the key down about three times as long as for a dot. You have sent the letter "T." Send one dash, pause a moment, send three dashes, pause a moment, and send two dashes. In so doing you have spelled out the name "TOM."

2. Now send three dots, pause a moment, send three dashes, pause a moment, and send three dots again. In this sequence you have spelled out "SOS." This is the international distress signal. Whenever the "SOS" signal is transmitted, all stations stand by for the distress message. A nearby station will then communicate with the station in distress, and rescue operations will be started. When distress messages are sent, all stations stand by to determine if they can help with the rescue operation and to prevent interference with the distress messages.
3. Additional code groups are as follows:

```
A  · —          N  — ·          C  — · — ·      Q  — — · —
U  · · —        D  — · ·        F  · · — ·      Y  — · — —
V  · · · —      B  — · · ·      L  · — · ·      Z  — — · ·
W  · — —        G  — — ·        P  · — — ·      X  — · · —
J  · — — —      K  — · —
                R  · — ·
```

Memorize these groups as best you can. After you have covered all the groups, practice your code by sending the following group of five-letter words:

THEME	ORDER	ZEBRA	EXERT	CHECK
LISTS	CRASH	QUEER	VILLA	UNTIL
TEXTS	DRUGS	OMITS	YEARS	JOKER
GRINS	BRICK	NEVER	WATER	ANGLE
FAINT	KEEPS	POINT	XRAYS	POLES

When you take a novice license examination, the code will be sent to you in a similar group of five-letter words. Twenty-five of these words will be sent, and if you can copy any successive five words correctly, you pass the code examination.

4. It is very helpful in learning the code to hear how properly sent code should sound. In fact, most authorities recommend that you learn each letter as a sound rather than as a combination of dots and dashes. A number of recorded code courses are available for this purpose, such as the *International Code and Training System* produced by Howard W. Sams & Co., Inc.

6

Radio-Frequency Oscillator

There are are many applications for oscillators in electronics. Most transmitted radio waves originate as a weak signal generated by a simple tube or transistor oscillator. Additional transistor and tube amplifier stages build up this weak signal to thousands and even hundreds of thousands of watts of transmitted radio-frequency power.

There are oscillators used in your local radio station. One of these oscillators generates a very weak signal on the frequency of your local broadcast station. If your local station operates on 610 kHz, a highly stable oscillator is used to generate this signal. Additional radio-frequency amplifiers then increase the power of this weak oscillation to the rated power of your local broadcast station. Even your radio receiver that picks up the station very probably uses a radio-frequency oscillator.

Oscillators are also used in two-way radio equipment, radio-controlled models, and even electronic garage-door openers. Electronic test equipment uses a variety of oscillators. If you know a radio amateur, chances are he also uses a variety of radio-frequency oscillators in his station. From these examples it is apparent that the radio-frequency oscillator is important to many electronic systems and devices.

RADIO-FREQUENCY FEEDBACK

The radio-frequency oscillator does not differ greatly from the audio-frequency oscillators discussed in the previous chapter. Radio-frequency oscillators have one or more resonant circuits. In fact, the frequency of oscillation is determined largely by the characteristics of the associated resonant circuit. The resonant circuit is so arranged in the stage that a portion of the radio-frequency energy in the resonant circuit is fed back to the input circuit. In this manner oscillations are sustained as explained in connection with an audio oscillator.

The oscillator shown in Fig. 6-1 is called a Hartley type. In this oscillator, the inductor of the resonant circuit is tapped. In the example of Fig. 6-1, the tap is placed at ground potential. The collector circuit is connected to one side of the resonant circuit; the base circuit is connected to the other. In so doing, the rf voltage between the base side of the tank circuit and ground is essentially out of phase with the rf voltage between the collector side of the resonant circuit and the ground tap. The voltage between the base side of the resonant circuit and the tap is the feedback component. This rf voltage, which is fed back to the base via blocking capacitor C1, is in phase with the base-emitter voltage.

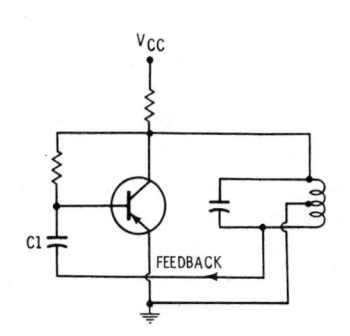

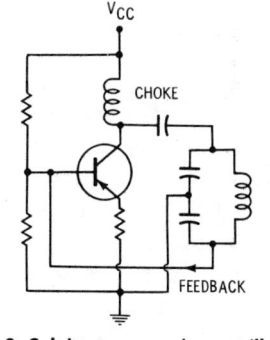

Fig. 6-1. Hartley-type transistor oscillator.   Fig. 6-2. Colpitts-type transistor oscillator.

As you learned, the collector voltage and the base voltage for the transistor stage are 180° related. Hence, the tank-circuit arrangement provides the additional 180° phase shift needed to obtain positive feedback, a necessary condition for circuit oscillation.

As in the case of the audio oscillators discussed in the last chapter, the oscillator is self-excited. Some slight signal variation starts the cycle of events that builds up to a self-sustaining oscillation. The build-up of feedback voltage is greatest at the resonant frequency of the tank circuit. As a result, the characteristics of the resonant circuit determine the frequency of oscillation. As with transistor-amplifier operation, the collector-base junction is reverse-biased, and the base-emitter junction is forward-biased.

RF OSCILLATOR TYPES

Some of the most popular radio-frequency oscillators are shown in Figs. 6-2 through 6-4. A transistor Colpitts oscillator is shown in Fig. 6-2. This oscillator is comparable to the Hartley type, except that a split capacitor feedback arrangement is used instead of a split inductor. In this case, the ground point is located at the junction of the resonant-circuit capacitors. The Colpitts is a popular, high-frequency oscillator because of its good stability.

In the oscillator circuit of Fig. 6-2 the base bias is obtained from a voltage-divider resistor arrangement. An emitter resistor is used to improve the stability of the circuit. The collector voltage can be applied through a resistor or a radio-frequency choke coil. The presence of the choke coil, which has a very high reactance at the oscillating frequency, prevents the battery from loading the oscillator and decreasing the strength of the oscillations. When there is too severe loading of an oscillator by a component part or the load that is supplied with the signal from the oscillator, the feedback can be reduced to the point at which oscillations cease.

Two popular transistorized oscillator types are shown in Fig. 6-3. They are referred to as tuned-base (Fig. 6-3A) and tuned-collector (Fig. 6-3B) oscillators. In the tuned-base circuit, the feedback voltage is obtained by mutual coupling between the inductor in the collector circuit and the secondary coil, which is a part of the base resonant circuit. The windings are so arranged that there is a 180° phase shift between the primary and secondary voltages. Therefore, the feedback voltage is put in phase with the base voltage, and positive feedback exists.

The supply voltage is fed through a resistor directly to the collector; the collector voltage is said to be *shunt-fed*. This means the supply voltage, through an isolating resistor R1, is connected in parallel, or in shunt, with the oscillator resonant circuit.

(A) Tuned-base oscillator.

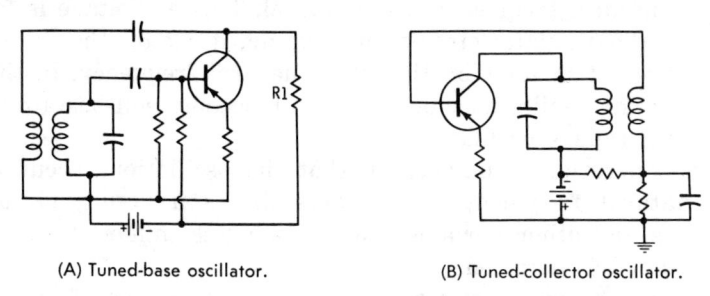

(B) Tuned-collector oscillator.

Fig. 6-3. Feedback-coil transistor oscillators.

It is also possible to use a tuned-collector arrangement as shown in Fig. 6-3B. In this case, the resonant circuit is in the collector lead, and the secondary coil is used as the feedback link between the inductor of the resonant circuit and the base of the transistor. Again, the winding arrangement is such that positive feedback is present. In this circuit, the supply voltage is connected to the bottom side of the collector tank circuit. In this case, the supply voltage is said to be *series-fed* because the battery and the resonant circuit are in series between the collector and emitter of the transistor.

Each transistorized oscillator type either has some special characteristic or can be best adapted to a particular electronic application. A crystal-controlled type of oscillator (Fig. 6-4) is highly stable. In transistor circuits, change in operating conditions, temperature, and values of the frequency-determining resonant circuit can affect the frequency of oscillation. Thus, the frequency of the oscillator output may not remain constant.

The advantage of a crystal oscillator is that the crystal itself has a high stability and at the same time displays the characteristics of a resonant circuit. A crystal is made of a crystalline material which can be made to vibrate physically with the application of an ac voltage. The vibration of the crystal occurs at its natural, mechanical resonant frequency. Furthermore, in the process of vibrating, the crystal also generates an ac voltage.

When a crystal is cut to vibrate at a specific frequency and is inserted in a transistorized oscillator circuit, it controls the frequency of oscillation. When any slight random radio-frequency variation occurs, the crystal goes into mechanical vibration. In turn, it generates an ac voltage that depends on the natural frequency of the crystal. This ac voltage is fed back in the oscillator circuit. In so doing, it excites the crystal to an even stronger vibration at its natural frequency. In this manner the oscillations build up and become self-sustaining as in any type of oscillator.

The operation is unusual in that the oscillations occur at the natural frequency of the crystal. As a result, circuit changes and other variables have much less influence on the frequency of operation.

The a-m broadcast stations in your area operate on frequencies that are crystal controlled. For example, if one of your local broadcast stations transmits on 1020 kHz, its frequency will be under the control of a crystal that has a natural resonant frequency of 1020 kHz.

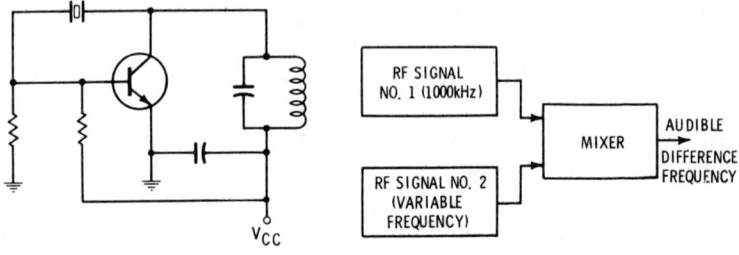

Fig. 6-4. Crystal-controlled transistor oscillator.

Fig. 6-5. Beating of two rf signals in a mixer.

ZERO-BEATING

In school or elsewhere, you may have learned about tuning forks. When they are struck, they vibrate mechanically at their natural frequency. You may have also listened to two tuning forks in vibration, one on one frequency and the other on a different frequency. If you listened carefully, you not only heard the two basic tones emitted by the vibrating forks, but you may have heard a much lower tone and/or a much higher tone as well. This condition occurred when the two tones mixed in your ear. The one or two additional frequencies heard are a sum frequency and a difference frequency. If one fork vibrated at 1000 Hz and the other at 800 Hz, it is very possible that you could have heard a lower-frequency, 200-Hz tone (1000 − 800) and/or a higher-frequency, 1800-Hz tone (1000 + 800).

Mixing of this type can also take place in electronic circuits. When two radio frequencies are brought very near to each other in a mixing circuit so that their difference in frequency is relatively low, you are able to hear an audible tone. For example, assume that the signal from a 1000-kHz radio-frequency oscillator is applied to a mixer (Fig. 6-5). If you now apply a signal from another oscillator that operates on 1001 kHz and tune in on the output of the mixer, you will be able to hear a 1000-Hz tone as an audible note. This 1000-Hz note is the difference between the two radio-frequency oscillator frequencies (1,001,000 − 1,000,000).

Now go a step further and vary the frequency of one of the oscillators in relation to the other. If you set the second oscillator on a frequency of 1005 kHz, the difference frequency is 5000 Hz, and a relatively high-pitched audio tone is heard. When you gradually decrease the frequency of the second oscillator, the frequency of the audible tone also decreases correspondingly. When the frequency of the variable oscillator is only 60 Hz higher than the fixed-oscillator frequency, you hear a 60-Hz tone.

Furthermore, if you continue to bring the frequency of the variable oscillator nearer and nearer to the frequency of the fixed oscillator, the difference frequency between the two oscillators will eventually become zero, and you will hear no tone output.

If you now continue to lower the frequency of the variable oscillator, its frequency will be on the low-frequency side of the fixed-oscillator frequency. Nevertheless, there will be a difference present, and the tone will become audible. In fact, the frequency of the audible tone will begin to increase as the frequency of the variable oscillator is lowered. When the variable-frequency oscillator is set on a frequency of 995 kHz, a 5000-Hz note will again be heard.

The process just described is referred to as *beating* the two signals together. When the two oscillators are at exactly the same frequency, there will be no audible output. This is referred to as the *zero-beat* point. If the frequency of the variable oscillator is moved in either direction from this zero-beat point, the frequency of the audible tone will increase.

The fact that you can hear the increase in frequency on each side of a definite setting is a help in many electronic measurements. For example, if you wish to check the frequency of a given oscillator, you can do so by using a second oscillator that has a variable, but always known, frequency. When you cause the variable-frequency oscillator to zero-beat with the unknown frequency, you know that the frequency of the variable oscillator has been set at the exact frequency of the oscillator being checked. If the dial of your variable oscillator is calibrated, you can then say that the previously unknown frequency is so many kilohertz as read on the calibrated dial. In this case, you are using the calibrated variable oscillator as a *frequency meter*. In the demonstrations that follow, you will be able to hear the zero-beat between a transistor oscillator that you will build, and a signal from a local radio station. The parts list in Table 6-1 includes the projects for Chapters 6 and 7.

DEMONSTRATION 1
FET RF Oscillator

The field-effect transistor performs well in oscillator circuits. Simple, easy-starting arrangements are possible. The crystal radio circuit of the Chapter 4 experiment can be rearranged into the oscillator circuit of Fig. 6-6. This is known as an Armstrong feedback oscillator and is quite popular in

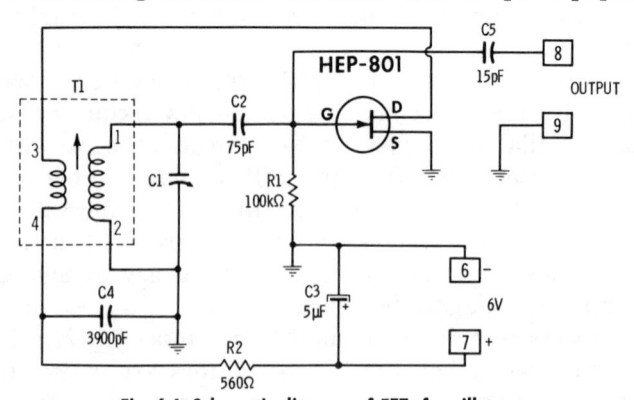

Fig. 6-6. Schematic diagram of FET rf oscillator.

superheterodyne receivers. A gate resonant circuit is employed using the antenna coil transformer that was a part of the crystal detector circuit in Chapter 4. The feedback which sustains oscillations is coupled from the drain of the transistor by way of the second winding of the transformer. Inductive coupling serves as a feedback path to the gate resonant circuit. Oscillations can be removed at the gate and coupled through a low-value capacitor (C5) to the output clip.

Table 6-1. Parts List for Chapters 6 and 7

Quantity	Description
1	Pegboard, 12 in. by 8 in.
12	Fahnestock clips.
1	Lantern battery, 12 V.
1	Broadcast antenna coil, universal replacement (Stancor RTC-8736 or equiv.).
1	Field-effect transistor (HEP-801).
1	Variable capacitor, 365 pF.
1	Vernier logging dial, with knob (calibrated 0 to 100 in 180° rotation).
1	Capacitor, 5 μF, 15 V.
1	Capacitor, 3900 pF.
1	Capacitor, 75 pF.
1	Capacitor, 15 pF.
1	Resistor, 100 kΩ, ½ W.
1	Resistor, 560 Ω, ½ W.
1	Resistor, 270 Ω, ½ W.

Procedure

1. First, remove the crystal detector circuit from the pegboard used in Chapter 4. Most of the clips can remain in their same positions and there is no necessity for changing the clips and lugs associated with the FET circuit. The wiring and parts values are changed however. The two clips associated with the antenna and ground of the crystal radio of Chapter 4 are now moved to the right side of transformer T1 as shown in the photograph and pictorial diagram of Fig. 6-7.
2. Connect the lead from terminal 1 of the transformer to clip 1; connect the lead from terminal 2 to ground clip 5. Connect the stator of variable capacitor C1 to clip 1; connect

its rotor to clip 5. Connect a jumper from clip 5 to output clip 9.
3. Connect the lead from terminal 3 of the transformer to clip 2; connect the lead from terminal 4 to clip 3. Connect the drain clip of the FET to clip 2. This closes the feedback path.

(A) Photograph of pegboard assembly.

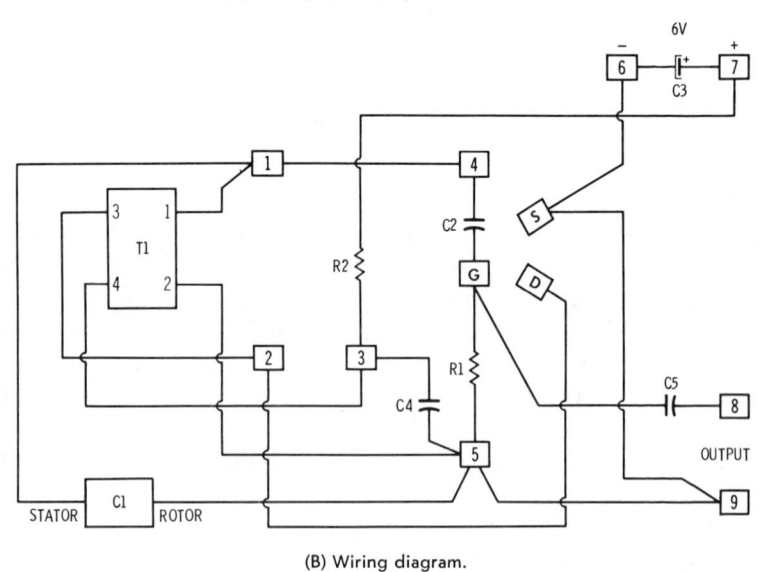

(B) Wiring diagram.

Fig. 6-7. FET rf oscillator demonstration.

Connect decoupling resistor R2 from clip 3 to battery clip 7. Connect decoupling filter capacitor C4 between clip 3 and clip 5. This completes the connection of the drain and feedback circuits.
4. Connect the source clip of the FET to battery clip 6. Also connect the source clip to output clip 9. Connect filter capacitor C3 between clips 6 and 7.
5. Complete the gate circuit by connecting a jumper between clips 1 and 4; connect capacitor C2 between clip 4 and the gate clip. Connect resistor R1 between the gate clip and ground clip 5. Connect output capacitor C5 between the gate clip and output clip 8.
6. The oscillations can be received on any a-m radio placed near clip 8. In fact, for good reception, a piece of insulated wire can be connected to clip 8 and then draped loosely across the radio. Tune the radio to some broadcast station between 540 and 800 kHz. Vary oscillator capacitor C1 very slowly from its fully meshed position to where a strong beat note (whistle) can be heard on the radio. As you continue to vary the capacitor onward, the beat note will become lower and lower in frequency.
7. Tune capacitor C1 very carefully now to obtain an exact zero beat (no tone) on the received signal. If you have obtained zero-beat, it means that the tone will be heard as you vary the capacitor on either side of its exact zero-beat (no tone) setting. The frequency of the oscillator is now exactly the same as the frequency of the broadcast station being picked up on the radio.
8. Tune in a station somewhere between 540 and 620 kHz. The capacitor plates should be almost fully meshed. If not, it is possible to retune the adjustment of transformer T1 until the lowest-frequency broadcast station is received with capacitor plates almost fully meshed.

DEMONSTRATION 2
Frequency Measurement and Calibration

Frequency is an all-important consideration in many electronic systems. What is the transmitting frequency of a given station? What is the resonant frequency of a tuned circuit? To what frequency is a receiver tuned? What is the frequency of the oscillator? Over what frequency range can you adjust the oscillator? Over what frequency range will a receiver tune or a transmitter operate?

Introduction

Certain basic techniques are used to measure frequency and/or set a given oscillator or transmitter on a specific frequency. In this demonstration, the zero-beat technique will be used to set the rf oscillator on a specific frequency.

Procedure

1. Turn on the oscillator and the a-m receiver. Tune in a strong broadcast station that operates on any frequency between 540 and 650 kHz.
2. Vary the frequency of the rf oscillator until a beat note (sometimes called a heterodyning tone or whistle) is heard. If more than one heterodyne can be heard as the oscillator capacitor is tuned over its range, choose the strongest.
3. Try to zero-beat the tone. In so doing, you have set the rf oscillator on the same frequency as the received station signal.
4. Note the dial setting for this matching of oscillator and broadcast-station frequency. It is a rather accurate reading of the oscillator frequency because broadcast stations must maintain their carrier frequencies between ±20 Hz of the assigned value. Note the dial setting in Table 6-2, along with the frequency assigned to the particular broadcast station.
5. Tune in a broadcast station that operates somewhere between 650 and 800 kHz. Repeat the zero-beat technique. Again note the dial setting and precise frequency of the

Table 6-2. Form for Frequency-Calibration Data

	Station	Frequency	Dial Setting
1.			
2.			
3.			
4.			
5.			
6.			
7.			

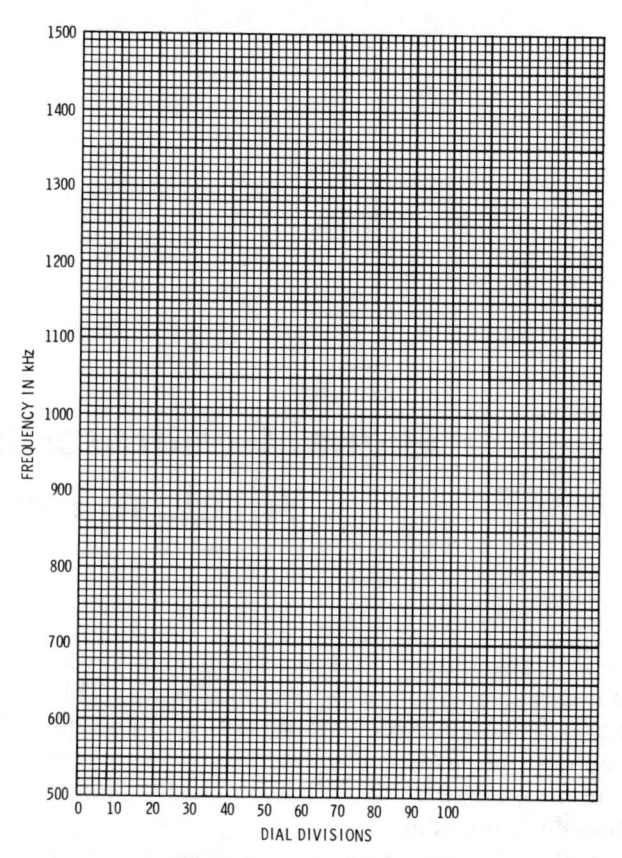

[Graph: Frequency in kHz (500–1500) vs Dial Divisions (0–100)]

Fig. 6-8. Frequency vs dial setting.

broadcast station in the chart. Repeat this procedure for a broadcast station in each of the following ranges: 800-950 kHz, 950-1100 kHz, 1100-1250 kHz, 1250-1400 kHz, and 1400-1550 kHz.

6. You can now plot a curve of frequency versus dial setting on the graph of Fig. 6-8. Locate each of the points you obtained in preparing the chart. Draw a curve connecting all of the points. You will now have a plot of dial setting versus frequency for your rf oscillator. It is now possible to set your oscillator on almost any frequency by setting the capacitor to the appropriate dial setting for the desired frequency.

7

Signal Generator and Wireless Player

In radio transmission, it is necessary to impress the voice or music signal on the radio-frequency wave. This is called *modulation*. Transistor oscillators, such as those described in Chapter 6, can be readily modulated by voice or music. In the project of this chapter you will modulate the transistor oscillator built in Chapter 6.

It is customary, for economy and frequency-stability reasons, to modulate the amplifiers instead of the oscillators in commercial equipment. Thus, the modulation does not occur until the radio-frequency signal has been amplified to a considerably higher power level. Nevertheless, there are a few commercial applications that do employ direct modulation of an oscillator.

There are various types of modulation. This chapter is concerned only with *amplitude modulation* or *a-m*. In this modulation process, the amplitude of the radio-frequency wave is made to vary with the variations of voice or music. The first step in modulating an rf wave is to convert the speech or music into a corresponding electrical variation (voltage or current). So-called *transducers* are used to make the conversion. These were covered in Chapter 3.

AMPLITUDE MODULATION

In Chapter 6 you learned that the limit of the oscillatory build-up is set by various circuit factors, which include the biasing of the transistor junctions. In addition, by changing the junction biasing, it is possible to control the amplitude which the oscillations will reach. By changing the base-emitter bias or the collector-base bias, it is possible to regulate the amplitude of the rf oscillations formed.

In the amplitude-modulation process, the bias of one or both junctions is changed at the voice or music frequency by the modulating circuit. As a result, the amplitude of the rf oscillations present in the resonant circuit will also vary with the modulating wave.

Fig. 7-1 shows the influence of a change in base bias on the amplitude of the oscillations in the rf resonant circuit. Over the normal operating range, the higher the forward base bias, the greater is the magnitude of the rf signal in the resonant circuit. A low forward bias results in a weaker rf signal in the resonant circuit.

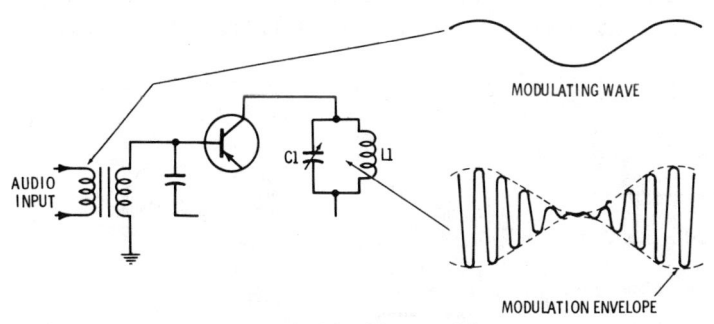

Fig. 7-1. Principle of base modulation.

If you vary the base bias with an applied audio sine wave as shown, the radio-frequency signal will also vary in magnitude, following the base-bias change. The resultant amplitude variation of the radio-frequency signal is referred to as an amplitude-modulation *envelope*. Note that the modulating wave itself does not appear in the resonant circuit. Nevertheless, its makeup is represented by the changes in magnitude of the rf oscillations. In summary, in the modulation process the amplitude of the oscillations has been varied by causing the base-emitter bias to vary with the modulating wave.

When there is a substantial variation in the base bias, there is a significant variation in the magnitude of the rf signal. If only a weak signal variation is applied to the base, there is a correspondingly smaller variation in the modulation envelope. Thus by controlling the amplitude of the modulating wave at the base, the modulation envelope can be made to follow changes in the level of voice or music. In effect, the modulation envelope follows both the frequency and the changes of amplitude of a modulating signal.

Some typical modulation-envelope examples are shown in Fig. 7-2. Figs. 7-2A through D show the influence of the frequency and amplitude of a modulating sine wave on the envelope variations. Actual voice and music variations are very random and indefinite, and appear more like Figs. 7-2E and F.

Examples of weak and strong modulation are given in Figs. 7-2E and F. Normally the level of the modulating wave and the operating conditions are chosen so that there is ample modulation of the rf wave. This represents the most efficient method of radio-wave transmission, delivering a stronger demodulated output at the receiver. Of course, the actual strength of

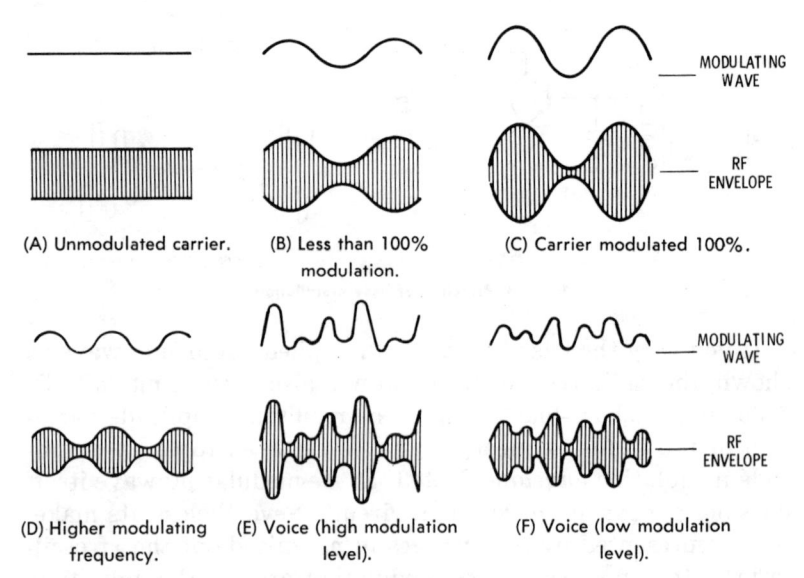

(A) Unmodulated carrier.
(B) Less than 100% modulation.
(C) Carrier modulated 100%.
(D) Higher modulating frequency.
(E) Voice (high modulation level).
(F) Voice (low modulation level).

Fig. 7-2. Modulation envelopes.

modulation also varies with the type of information to be transmitted. The modulation level will be substantially weaker in the transmission of a quiet piano passage as compared to the modulation level caused by a blaring trumpet.

Modulation level is often given as a percentage. An example of 100% modulation is shown in Fig. 7-2C. For 100% modulation, the magnitudes of the envelope cycles vary between zero and twice the magnitude of the unmodulated rf carrier. A lower modulation percentage exists in Fig. 7-2B.

TRANSISTOR MODULATION METHODS

In the example in Fig. 7-1, the modulating signal was applied to the base of the transistor. Two other methods of modulating a transistor oscillator are shown in Fig. 7-3. In Fig 7-3A, an audio transformer is placed in the path that connects the dc voltage to the collector. When an audio signal is present across the primary of the so-called *modulation transformer*, there is a similar change in the secondary voltage that adds to and subtracts from the dc collector voltage. In effect, it is causing the collector-base bias to vary.

The collector voltage also influences the magnitude of the rf signal. When the collector voltage is high, a high-amplitude rf signal results. With a low collector voltage, the magnitude of the rf signal is less. By varying the collector voltage with a modulating wave, there is a corresponding variation in the magnitude of the rf signal. Again an amplitude-modulated envelope is produced.

In general, a higher-amplitude modulating wave is needed to modulate the collector than is necessary at the base for a given modulation percentage. The application of the modulat-

(A) Collector modulation.   (B) Emitter modulation.

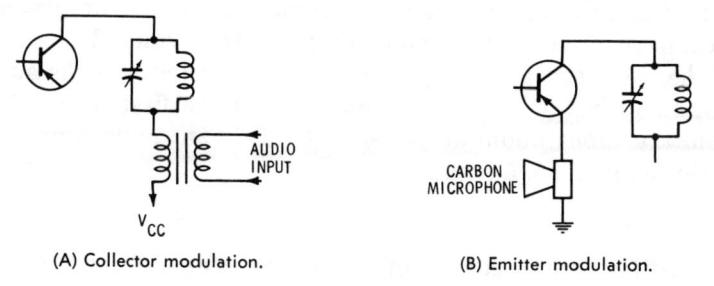

Fig. 7-3. Transistor modulation methods.

ing wave to the base also takes advantage of the amplification capability of the transistor stage. However, the advantages of collector modulation are that the modulation is more linear, and the envelope variations follow more closely the variations of the modulating wave.

The modulating wave can also be applied to the emitter, as shown in Fig. 7-3B. In effect, the biases of both junctions are then varied by the modulating wave. The input resistance in the emitter circuit is quite low, and this method of modulation is most suitable when the source of the modulating wave can operate into a low-resistance load. For example, carbon microphones have a low resistance and require a source of dc current for proper operation. Thus, one can be inserted between the emitter and ground, in which case the emitter current is made to vary when the microphone is activated by sound waves.

DEMONSTRATION 1
Tone Modulation

The output of a transistor rf oscillator can be modulated by a modulating signal applied to one or both junctions. The modulating signal, voice or music, is made to vary the junction bias at an audio rate. Since the junction bias also determines the magnitude of the rf signal developed across the resonant circuit, any change in the junction bias results in a corresponding change in the amplitude of the rf signal. The variations of the modulation envelope follow the modulating signal.

A FET oscillator can be modulated in a similar manner. The modulating wave can be applied to drain, source, or gate. In this experiment, the modulating wave is applied to the source, as shown in Fig. 7-4. The source input resistance is quite low and, therefore, reasonable match is made to the low-impedance secondary of a transistor audio-output transformer. The source of the audio signal for modulating the rf oscillator is the audio amplifier shown in Figs. 3-8 and 3-9. In the first demonstration, the input audio amplifier is used as an audio tone generator as in Figs. 5-5 and 5-6.

Procedure

1. Modify the rf oscillator of Fig. 6-6 by connecting a 270-ohm resistor (R3 in Fig. 7-4) between source and ground.

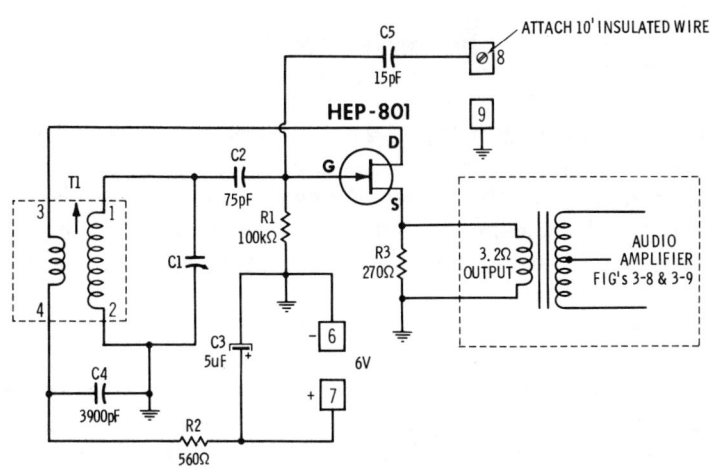

Fig. 7-4. Schematic of modulated rf oscillator demonstration.

2. Disconnect the loudspeaker of the audio amplifier and connect the secondary of the output transformer across source resistor R3.
3. Modify the input stage of the audio amplifier into the tone oscillator circuit of Figs. 5-5 and 5-6.
4. Turn on the rf oscillator, the audio amplifier, and the radio. Set the radio to an unoccupied frequency in the vicinity of 600-700 kHz. Adjust the frequency of the oscillator with variable capacitor C1 until a strong tone signal is received.

Many signal generators are similar to the tone-modulated oscillator arrangement you have just completed. When the oscillator output is tone modulated, it is no longer necessary to employ the zero-beat technique of the previous chapter. The tone can be heard whether or not the oscillator is set to the frequency of a received broadcast signal.

5. Reset the radio to some other unoccupied frequency and retune the oscillator until the tone can again be heard. It should be noted that with tone modulation, the transmitted signal is quite broad and it is more difficult to set the oscillator to some precise frequency than with the zero-beat technique. However, when using the zero-beat technique, it is necessary to have some other received comparison signal such as a broadcast signal. Tone modulation is more convenient in setting the rf oscillator to some unoccupied frequency on the broadcast band.

DEMONSTRATION 2
Wireless Player

The arrangement of Fig. 7-4 can be used as a wireless microphone or a wireless record player. In this application, the frequency of the rf oscillator is set to some unoccupied frequency in the broadcast band. Either a microphone or a phono cartridge can be used to supply signal to the input of the audio amplifier.

Procedure

1. Disconnect the tone oscillator circuit of the input transistor of the audio amplifier Using the input transistor as the first audio amplifier (Figs. 3-8 and 3-9), restore the audio amplifier circuit to normal.
2. Leave the loudspeaker disconnected and the secondary of the audio output transformer connected across resistor R3 in the source circuit of the rf oscillator.
3. Turn on the rf oscillator, the audio amplifier, and the radio. Set the oscillator frequency to some unoccupied frequency on the broadcast band between 550-700 kHz.
4. Apply signal from either a microphone or a phono cartridge to the input of the audio amplifier. Tune the receiver carefully for the best output. If there is some distortion, reduce the audio amplifier gain to a point below the distortion level. A strong, good-quality output can be obtained within about 30 feet of the rf oscillator. A stronger signal can be received by attaching about ten feet of insulated wire to the output of the rf oscillator to serve as a simple antenna.

8

Regenerative Receiver and Integrated-Circuit Audio Amplifier

Receiver sensitivity and output can be increased with the use of a more sensitive detector and an audio amplifier with higher power gain. A high output can be obtained in an output stage that uses an integrated circuit (IC). A further performance improvement is possible by using a detector that is capable of amplifying the incoming signal. An additional signal boost is given with the use of a controlled amount of positive feedback, which is referred to as *regeneration*.

TRANSISTOR DETECTORS

Two transistor circuit arrangements that can be used to demodulate an incoming a-m carrier are shown in Fig. 8-1. As you learned in Chapter 4, in the demodulation process only one side of the modulation envelope is passed; that is, the a-m signal is rectified. This can be accomplished by correct biasing of the base-emitter junction and the collector-base junction.

In the detector of Fig. 8-1A, the value of the R1-R2 voltage divider is selected so that the base-emitter junction is given a large forward bias. Thus, a high dc collector current flows. In this case, the negative variations of the modulation envelope do not cause any substantial change in collector current. However, the positive variations of the envelope decrease the forward bias of the junction and result in a like variation in the base and collector currents. The collector current variation is a replica of the variations of the modulation envelope. The advantage of this circuit arrangement is that the transistor provides signal amplification because there is a greater collector-current change than base-current change.

In Fig. 8-1B, the base-bias resistors and emitter resistor are of such value that the junction is biased nearly to cutoff (small

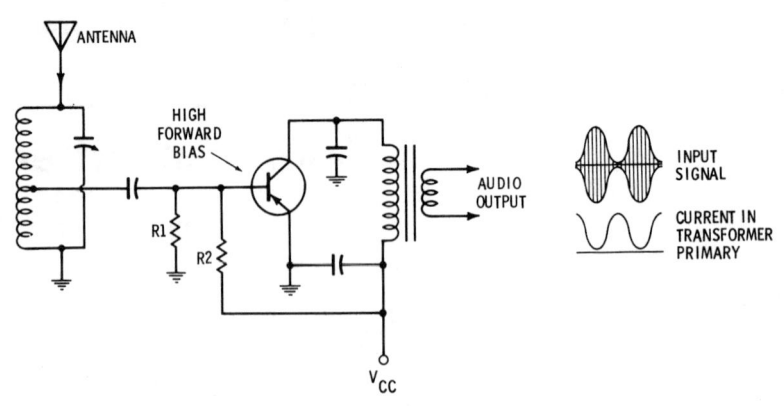

(A) Detector using high forward bias.

(B) Detector using low forward bias.

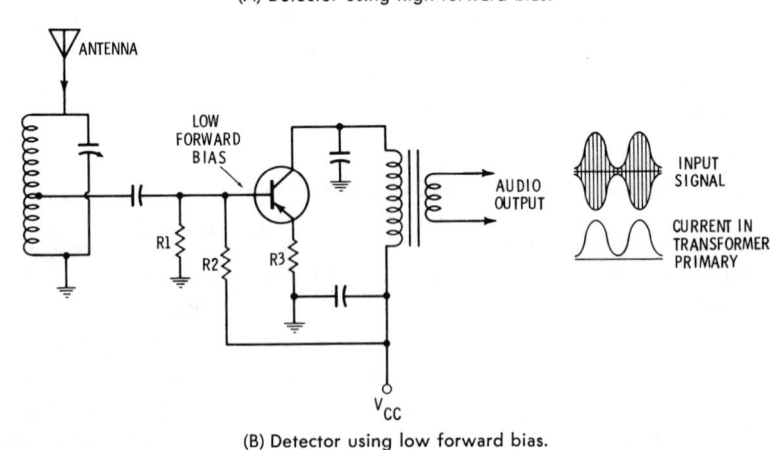

Fig. 8-1. Transistor detectors.

124

amount of forward bias). In this case, the positive alternations of the modulation envelope cause little change in collector current because they swing the base-emitter junction into the cutoff region. However, the negative variations of the envelope increase the forward bias and cause a like change in the base current. In turn, the base-current variations produce an amplified collector-current change. With a reasonably strong applied modulation envelope, this type of detector produces a strong demodulated output in its collector circuit.

In comparison to the circuit of Fig. 8-1A, the dc or no-signal collector current for the circuit of Fig. 8-1B is substantially lower. That is, circuit B is biased near cutoff, while circuit A has a high forward bias and, therefore, a higher no-signal collector current.

REGENERATIVE DETECTOR

In your work with audio- and radio-frequency oscillators, you learned about positive feedback. When there is adequate positive feedback, a small variation is built up until sustained oscillations are established.

Positive feedback can also be used to provide boost to an incoming radio-frequency signal. In so doing, the detector is made more sensitive. The signal boost is, of course, at a maximum at the incoming signal frequency because the associated tuned circuit is set to this frequency.

In the demodulation of an incoming a-m signal, the amount of feedback is controlled so that it is not great enough to produce self-sustained oscillations. Thus, positive feedback is inserted just up to the point at which a circuit goes into oscillation.

Regenerative Detector Circuit

A typical transistor regenerative detector is shown in Fig. 8-2. Basically, the circuit operates as a conventional transistor detector. Notice, however, there is a small coil (L2) which is coupled to the coil of the resonant circuit of the detector. Thus, a small amount of radio-frequency energy is fed back from the collector to the base circuit. Its level can be controlled by using capacitor C2. This feedback component is of such phase that it adds to the signal present in the resonant circuit; hence, the modulation envelope in the base resonant circuit is reinforced. As a result, there is a stronger demodulated output from the detector.

The amount of signal present in the feedback coil (referred to as a *tickler coil*) and the degree of coupling between the two coils is so regulated that there is maximum amplification without the presence of oscillations.

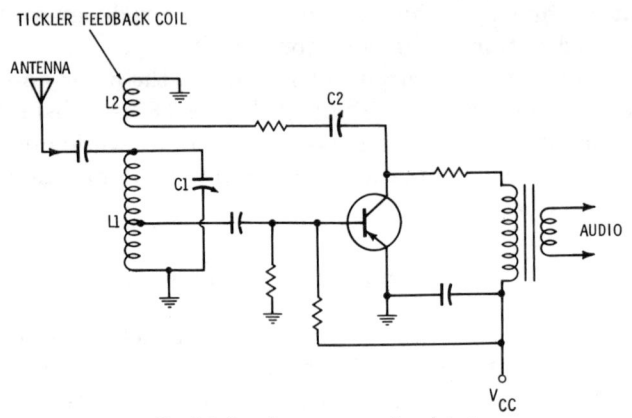

Fig. 8-2. Transistor regenerative detector.

The feedback must be positive, and, therefore, there must be a correct phase relationship between the rf components in the two coils. For example, if the tickler winding is reversed from the proper wiring, there will be negative feedback and a resultant decrease in the effective strength of the incoming signal. Thus, if your circuit tends to reduce the signal level, it is usually an indication that the two leads coming from the tickler coil should be reversed.

Code Reception

A regenerative detector can also be used to receive a code, or cw, signal, as shown in Fig. 8-3. In this operation, the positive feedback is permitted to build up to a level that will produce weak self-sustained oscillations. The oscillations occur at the fequency to which the base resonant circuit is tuned. To understand how these oscillations can recover the information from code transmissions, consider the make-up of a cw signal.

In transmitting code, the radio-frequency carrier is interrupted according to the dots, dashes, and spaces of the coded

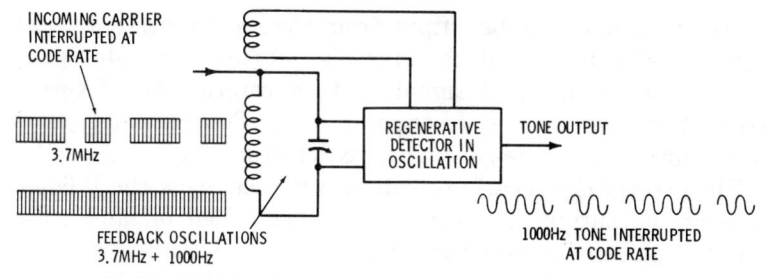

Fig. 8-3. Principle of code reception with a regenerative detector.

message. When transmitting a dot, the carrier is transmitted for a very short period of time; for a dash the carrier is on approximately three times as long. During the spaces between the dots and dashes, letters, and words, the carrier is turned off, and no signal is transmitted (Fig. 8-3).

The transmitted signal is an interrupted radio-frequency carrier. Inasmuch as the transmitted signal has a radio frequency, it cannot be heard when picked up by a conventional detector. The radio-frequency variations are far above the audible frequencies. How then can the interruptions of the rf carrier be reduced to an audible tone?

As you learned in Chapter 6, two radio-frequency carriers can be mixed, or beat, together to produce an audible difference frequency, or tone. If two radio-frequency signals are 1000 Hz apart, it is possible to derive a 1000-Hz audible tone by appropriate mixing. This is exactly the process that occurs when the feedback of a regenerative detector is increased to the level that produces self-sustained oscillations. These oscillations are beat against the incoming interrupted rf carrier to produce an audible output.

For example, the self-sustained oscillation of the regenerative detector (Fig. 8-3) can be set to 3.7 MHz plus 1000 Hz. If you now tune in a code signal on a frequency of 3.7 MHz, a 1000-Hz difference note will be present at the output of the regenerative detector. However, it is important to realize that the only time the tone is present is when there are two rf signals present in the base circuit. The feedback oscillations are always present, but the incoming rf carrier is interrupted in accordance with the coded message. Hence, the only times there are two radio-frequency signals present in the resonant circuit are when either a dot or a dash is being transmitted.

Thus, there will only be output from the regenerative detector during these dot and dash intervals. During the code spaces there is no incoming rf signal, and, therefore, two rf signals are not present in the resonant circuit. Consequently, there is no output from the regenerative detector.

The preceding operating conditions indicate that the 1000-Hz tone at the output of the detector is switched on and off in exactly the same manner that the transmitted radio-frequency carrier was turned on and off at the station. Hence, the audible tone at the output of the regenerative detector is a copy of the code message from the transmitting station.

The frequency of the output tone can be made 500 Hz, 2000 Hz, or any audio frequency by slight tuning of the regenerative-detector resonant circuit. You can set the self-sustained oscillations at the frequency that will produce the most favorable tone output for copying the cw message.

Table 8-1. Parts List for Chapter 8

Quantity	Description
1	Pegboard, 12 in. by 8 in.
16	Fahnestock clips.
1	Lantern battery, 12 V.
1	Tube socket, 5-prong.
4	Coil forms, 5-prong, 1½-in. dia. by 2-in. long (approx.).
1	Small spool, No. 20 enameled copper wire.
1	Small spool, No. 22 enameled copper wire.
1	Small spool, No. 26 enameled copper wire.
1	Radio-frequency choke, 2.5 mH.
1	Field-effect transistor (HEP-801 or equiv.).
1	Integrated circuit (HEP-580 or equiv.).
1	Potentiometer, 200 Ω.
1	Potentiometer, 15 kΩ.
1	Variable capacitor, 140 pF.
1	Variable capacitor, 15 pF.
2	Vernier dials, with knobs.
3	Capacitor, 10 µF.
2	Capacitor, 5 µF.
1	Capacitor, 510 pF.
1	Capacitor, 100 pF.
1	Resistor, 1.0 MΩ, ½ W.
1	Resistor, 330 kΩ, ½ W.
1	Resistor, 220 kΩ, ½ W.
1	Resistor, 100 kΩ, ½ W.
1	Resistor, 5600 Ω, ½ W.

DEMONSTRATION 1
FET Regenerative Detector

A regenerative detector has a high sensitivity. A single detector stage can produce enough output to operate a headset or drive an audio amplifier. The field-effect transistor has characteristics that make it adaptable for efficient use as a regenerative detector. It has a high-impedance input and places a light load on the input resonant transformer. This means efficient conversion from radio frequency to audio frequency and a reasonably good selectivity. It has good voltage gain and, therefore, a strong output voltage is developed which is reasonably free of distortion. Its high-frequency characteristics are such that it performs well in a regenerative circuit up to 20 MHz and higher.

Introduction

The parts needed for this chapter are listed in Table 8-1. The FET regenerative detector circuit for this project is shown in Fig. 8-4. Plug-in coils are used in order to cover a wide range of frequencies. There is a high-impedance resonant secondary and a low-impedance antenna input coil. These are wound on a five-prong plug-in coil form. A main tuning capacitor and a bandspread tuning capacitor permit fine tuning.

The feedback path is between drain and the common end of the resonant circuit. The source is connected to a tap on the secondary coil. The position of the tap determines the level of feedback. A potentiometer (R1) is connected across the coil

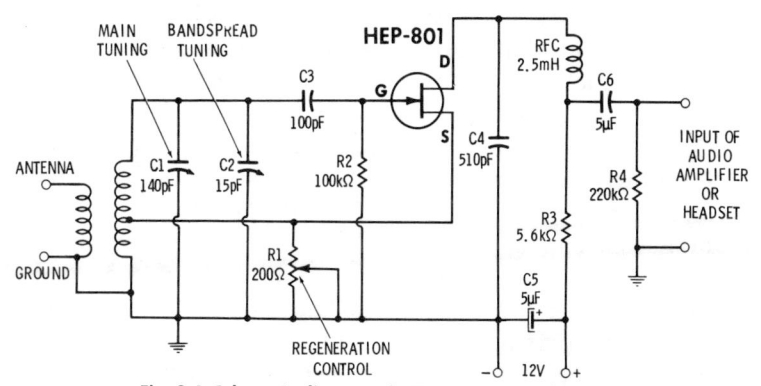

Fig. 8-4. Schematic diagram of FET regenerative detector.

between the tap and common. The setting of this potentiometer permits fine adjustment of the feedback level. The feedback level is set to produce self-oscillation when receiving cw or single-sideband signals. The regeneration control must be set slightly below the feedback point in the reception of a-m signals. However, the best selectivity in receiving even a-m signals is obtained with a small amount of self-oscillation. In this case, it is necessary to zero-beat the desired signal very carefully with the bandspread tuning control until the heterodyning whistle is zeroed out.

The demodulated audio is developed across the drain load resistor R3. A radio-frequency choke (RFC) isolates the radio-frequency and audio components. The output coupling network consists of capacitor C6 and resistor R4. In receiving a strong signal, there is enough output to drive a headset directly. The output signal can also be applied to the input of the audio amplifier constructed in Chapter 3 (Figs. 3-8 and 3-9). In this case, the signal is amplified to speaker level. In fact, speaker volume and audio quality are surprisingly good.

Procedure

1. The pegboard arrangement used in the previous two chapters is the basis for the construction of the regenerative detector. Remove the broadcast antenna coil and substitute a 5-prong tube socket (see Figs. 8-5 A, B, C, and D).
2. Mount the two variable capacitors. Mount the regeneration control to the right of the bandspread capacitor.
3. Four plug-in coils provide frequency coverage from the a-m broadcast band up to approximately 15 MHz. Coil 1 covers the high end of the broadcast band, the marine band, plus the 160-meter amateur band. Coil 2 covers marine and aviation frequencies as well as the 80-meter amateur band. Coil 3 covers the 40-meter band and 41- and 49-meter international short-wave bands. Coil 4 covers the 20-meter amateur band and the 10-, 25-, and 31-meter international broadcast bands.

   All coil construction data are given in Figs. 8-5C and D. For example, for band 4, the secondary winding mounts above the primary winding and consists of nine close-wound turns of No. 22 enameled copper wire. The tap is made at the third turn from the bottom. The primary winding is wound below the secondary and consists of four close-wound turns of No. 20 enameled copper wire. This coil can be seen lying on the pegboard in the lower right-hand corner of Fig. 8-5A.

In constructing the coils, it is wise to wind the primary first, positioning it at the bottom of the plug-in coil form. Connect the top of the primary winding to pin 4 of the coil form; connect the bottom of the primary to pin 3. The wire ends are inserted into the pins through small holes drilled in the coil form. Solder the wires at the tips of each pin (see Fig. 8-5D for more details).

Next, wind the secondary starting just above the primary after a connection is first made to pin 2. Wind the secondary until you reach the tap turn, twist the wire over a suitable length, insert it through a hole drilled in the coil form and insert through pin 1. Continue winding the secondary until the total number of turns is obtained. Cut the wire making sure there is enough added length to permit insertion into pin 5.

Note: Band 1 covers the high-frequency end of the broadcast band using the 140-pF and 15-pF variable capacitors. If you wish to cover the entire broadcast band, you can use the same coil. In this case, use a 365-pF variable capacitor for C1 and the 140-pF capacitor for the bandspread variable.

4. In wiring the socket for the coil form, pay particular attention to the terminal connections. At the base of the coil form the numbers are read clockwise. However, in looking down on the socket, the count-off must be made in a counterclockwise direction. This is shown in Fig. 8-5.
5. Connect the stators of the two variable capacitors together, then connect the two rotors together. Connect the stators to terminal 5 of the coil socket. Connect the rotors to input ground clip 1, and then connect clip 1 to terminal 3 of the coil socket. Connect terminals 2 and 3 of the coil socket together, and then connect them to battery clip 6. Connect terminal 4 of the coil socket to antenna clip 2.
6. Connect terminal 5 of the coil socket to clip 3. Connect clips 3 and 4 together. Connect terminal 1 of the coil socket to the source clip and also to one side of the regeneration control R1.

7. Connect capacitor C3 between clip 4 and the gate clip. Connect resistor R2 between the gate clip and ground clip 5. Join the opposite end of potentiometer R1 and the arm (center connection) to clip 5. Complete the common (ground) circuit by connecting clip 5 to clip 6 and to output clip 10.

(A) Photograph of pegboard assembly.

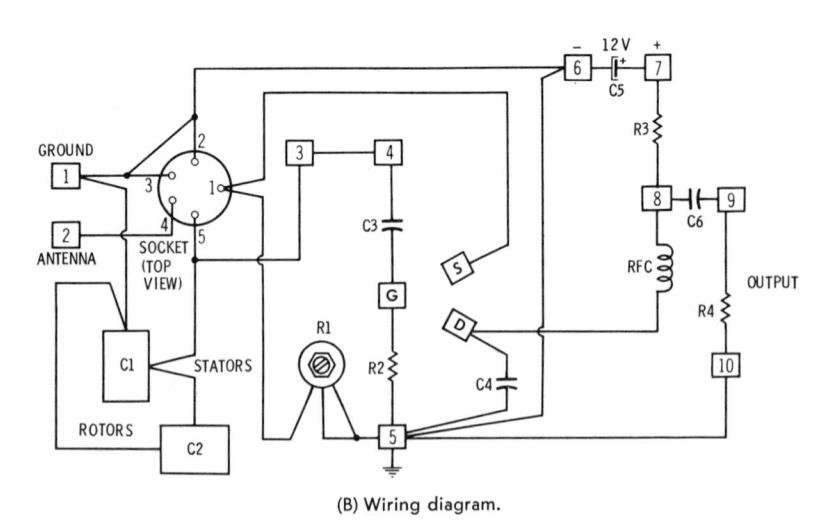

(B) Wiring diagram.

Fig. 8-5. FET regenerative

8. Connect the radio-frequency choke (RFC) between the drain clip of the FET and clip 8. Connect output load

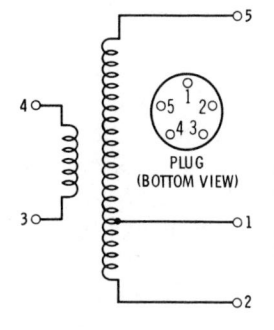

COIL	SECONDARY			PRIMARY	
	WIRE	TURNS	TAP FROM BOTTOM	WIRE	TURNS
1	#26	60	20	#20	5
2	#22	45	15	#20	5
3	#22	19	7	#20	4
4	#22	9	3	#20	4

(C) Coil construction data.

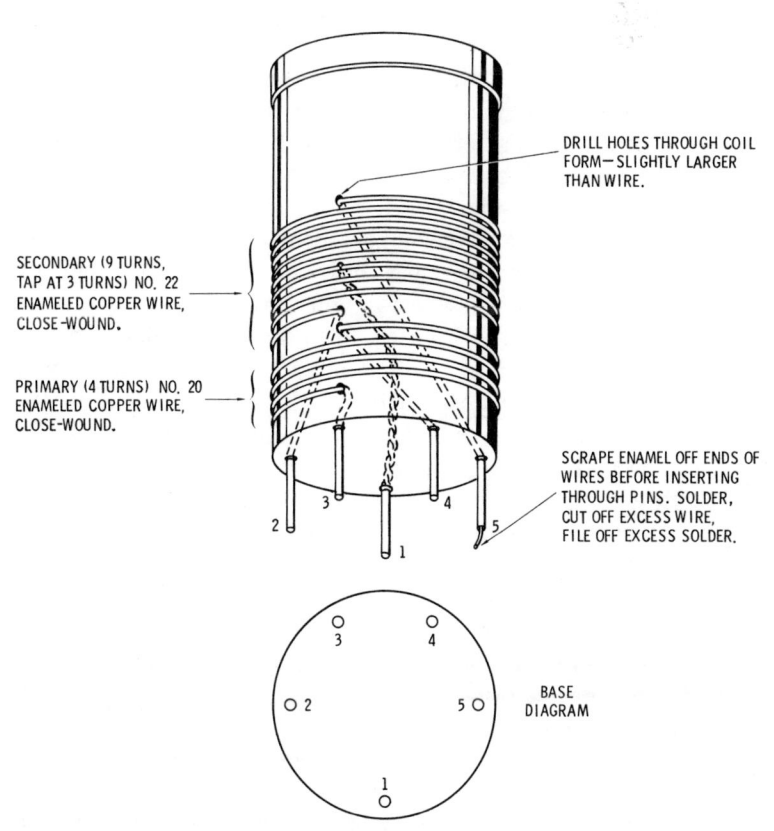

SECONDARY (9 TURNS, TAP AT 3 TURNS) NO. 22 ENAMELED COPPER WIRE, CLOSE-WOUND.

PRIMARY (4 TURNS) NO. 20 ENAMELED COPPER WIRE, CLOSE-WOUND.

DRILL HOLES THROUGH COIL FORM—SLIGHTLY LARGER THAN WIRE.

SCRAPE ENAMEL OFF ENDS OF WIRES BEFORE INSERTING THROUGH PINS. SOLDER, CUT OFF EXCESS WIRE, FILE OFF EXCESS SOLDER.

BASE DIAGRAM

(D) Construction details for coil D.

detector demonstration.

resistor R3 between clip 8 and battery clip 7. Connect capacitor C4 between the drain clip and clip 5.
9. Connect capacitor C6 between clips 8 and 9. Connect resistor R4 between clips 9 and 10. This completes the wiring of the regenerative detector.
10. Check your wiring carefully. Make certain it complies with the schematic diagram in Fig. 8-4. Connect two leads from the output of the detector to the input of the audio amplifier. Connect an antenna to clip 2 and a good outside ground to clip 1.
11. Plug in the band 1 coil. Set both variable capacitors for maximum capacitance (plates fully meshed).
12. Turn on the audio amplifier and detector. Set the regeneration potentiometer R1 for maximum resistance across the feedback portion of the secondary. Slowly decrease the capacitance of the main tuning capacitor. When you come to a station, you will hear the heterodyne whistle. As you keep tuning, the whistle will lower in frequency and finally zero-beat. At this setting you will hear the station best. If you continue decreasing the capacitance, the heterodyne whistle will again be heard.

Reset the capacitor for zero-beat on the received broadcast signal. Now decrease the resistance of the regeneration control very slowly. As you make the adjustment, at some point in the direction of minimum resistance there will be a buildup of signal and noise, and then a "plop" as you go from the oscillating to the nonoscillating setting. At this point, you will receive a strong signal and there will be no heterodyning whistle.

Best tuning results are always obtained near the point of changeover from the oscillating to the nonoscillating condition. For a-m reception, you usually can tune to near the point of changeover on the nonoscillating side. Strongest cw and single-sideband reception is obtained by setting the regeneration control just on the oscillating side of the changeover setting.

As you tune from one end of the band to the other, the changeover point shifts on the regeneration control. Thus, the regeneration control requires readjustment as you tune over the frequency range of each coil. The regeneration control also has some effect on receiver tuning.

DEMONSTRATION 2
Amateur and Short-Wave Radio Reception

The four coils constructed in Demonstration 1 cover four amateur radio bands and seven major international short-wave bands. The operation of the receiver will be similar for all four coils. As you go higher in frequency, however, you must make your tuning and regeneration-control adjustments more carefully. The presence of other objects, such as your hand or arm, near the coil and tuning capacitors will affect tuning, particularly on the high-frequency bands. This is the reason that vernier tuning knobs are recommended. The fine-tuning control is particularly useful in tuning high-frequency signals. Hand capacity can be further minimized by the use of an insulated shaft extension between the vernier tuning control and the shaft of the bandspread capacitor. This will keep your hand away from capacity-sensitive components, and will assure more accurate tuning.

Procedure

1. Insert band 1 coil in the socket. This coil tunes over the 160-meter amateur band between 1.8 and 2 MHz as well as the marine frequencies between 2 and 2.5 MHz. Aviation and weather stations can also be heard in this frequency range and above 2.5 MHz. Usually the signals on this frequency are stronger after dark and continue so until dawn. During the fall, winter, and spring months there is substantial nighttime activity on the 160-meter amateur band.
2. Insert band 2 coil in place of band 1 coil. Band 2 coil covers the 80-meter amateur band as well as the 60-meter (5 MHz) and 90-meter (3.3 MHz) medium-frequency broadcast bands. Some signals can be heard strongly during the day. Better results are obtained during the winter months rather than the summer months.

   Nighttime results are best, with considerable cw and phone activity on the 80-meter band. This band extends through 3.5 and 4 MHz. The phone operators are found between 3.8 and 4 MHz, while code signals are heard between 3.5 and 3.8 MHz. Slower-sending novice cw stations are heard between 3.7 and 3.75 MHz. This is a good band for code practice.

The 60- and 90-meter medium-frequency broadcast bands are crowded with signals during nighttime hours. In many parts of the country, it is also possible to pick up the WWV time signal which is transmitted on exactly 5 MHz.
3. Remove band 2 coil and insert band 3 coil. Its segment of spectrum includes the 40-meter amateur band. This band is active day and night. It also has cw, novice, and phone sections.

At nighttime, strong short-wave broadcast signals can be heard in a portion of the same frequency spectrum, the so-called 41-meter international short-wave band. You can also hear short-wave broadcast stations on the 49-meter band, (approximately 5 MHz). Day and night, one can usually find strong signals on this band.
4. Remove band 3 coil and insert band 4 coil. This spectrum includes the 20-meter amateur band. This band is also active day and night. Radio amateurs from all parts of the world can be heard transmitting on this frequency between 14 and 14.35 MHz. There are three active international broadcast bands. These are the 19-meter band at approximately 15 MHz, the 25-meter band at approximately 12 MHz, and the 31-meter band at approximately 10 MHz. Strong signals can be received during the day and up into the nighttime hours. The 25- and 31-meter international bands are active day and night with signals coming through from all parts of the world.

DEMONSTRATION 3
Integrated-Circuit Audio Amplifier

The integrated circuit (IC) has become an essential part of many radios, television sets, and other electronic equipment. Many tiny, almost microscopic components are mounted within a case the same size as, or not much larger than, an ordinary single transistor. Such a device has multiple leads so that external circuit connections can be made to the various components. There are some complex integrated circuits that have hundreds of individual discrete components.

Introduction

In this experiment, a relatively simple integrated circuit is used in the construction of a simple audio amplifier. Within the case of the HEP-580 integrated circuit there are four transistors and six resistors. There are eight external leads as shown in Fig. 8-6.

The transistors are connected in pairs with all emitters connected to a common terminal (lead 4). There are individual inputs for the four bases: leads 1, 2, 3, and 5. The collectors are paired and connected to leads 6 and 7. The 3600-ohm collector load resistors are joined and connected to the supply voltage terminal (lead 8).

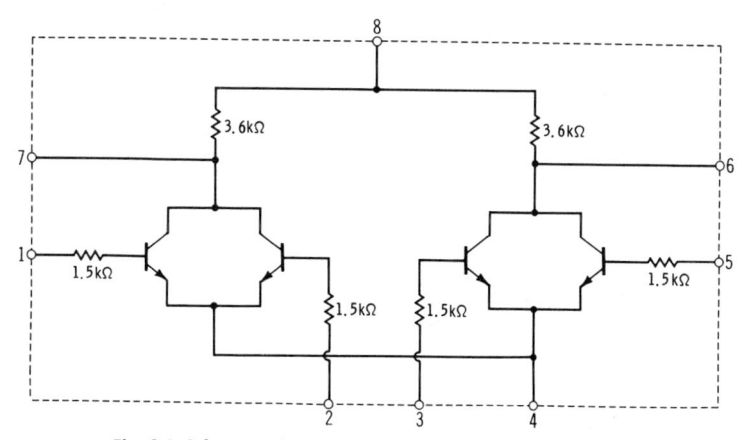

Fig. 8-6. Schematic diagram of the HEP-580 integrated circuit.

How this integrated circuit can be connected as a two-stage resistance-coupled audio amplifier is shown in Fig. 8-7. The input signal is supplied to leads 1 and 2, the input bases, by way of capacitor C6 in the regenerative detector circuit. The base-emitter junctions are biased with the 1-megohm base resistor. Capacitor C2 couples the collectors of the first transistor pair to the bases of the second transistor pair. Resistor R2 supplies biasing for the second transistor pair. Output is removed from the collectors of this pair at lead 6 and supplied through capacitor C3 to the headset.

By connecting a potentiometer (R3) between leads 8 and 7, the audio level can be controlled. Note that capacitor C2 is connected to the arm of the potentiometer. If the received signal is extremely strong, it is possible to remove an output at the collector of the first transistor pair, lead 7. Output from this point can be supplied to the input of the push-pull speaker amplifier constructed previously, permitting the amplification of very weak incoming signals.

Procedure

1. The actual wiring diagram of the IC audio amplifier is shown in Fig. 8-8B. The schematic is identical to that of Fig. 8-7 but shows the wiring with relation to the base leads of the HEP-580. The internal components are not shown. (*Note:* In Fig. 8-8B, the numbering of the HEP-580 leads is clockwise from the tab as viewed from the bottom of the unit. When viewed from the top of the unit, as in Fig. 8-8A, the numbering is reversed, or counterclockwise.)

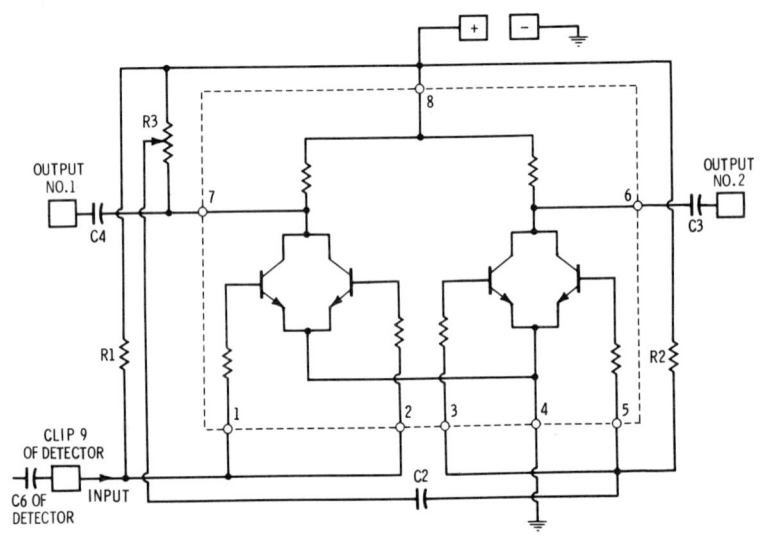

Fig. 8-7. The HEP-580 connected as a two-stage audio amplifier.

The integrated circuit is constructed by first mounting 8 small soldering lugs to the pegboard (see Fig. 8-8A). The external wiring is then completed, soldering leads to the individual lugs but making certain the lug holes are kept open for the mounting of the integrated circuit as the final step.

In soldering the individual leads of the IC, use long-nosed pliers as a heat sink as each individual lead is soldered to the appropriate soldering lug. The mounting arrangement can be seen in Fig. 8-8A.

(A) Photograph of pegboard assembly.

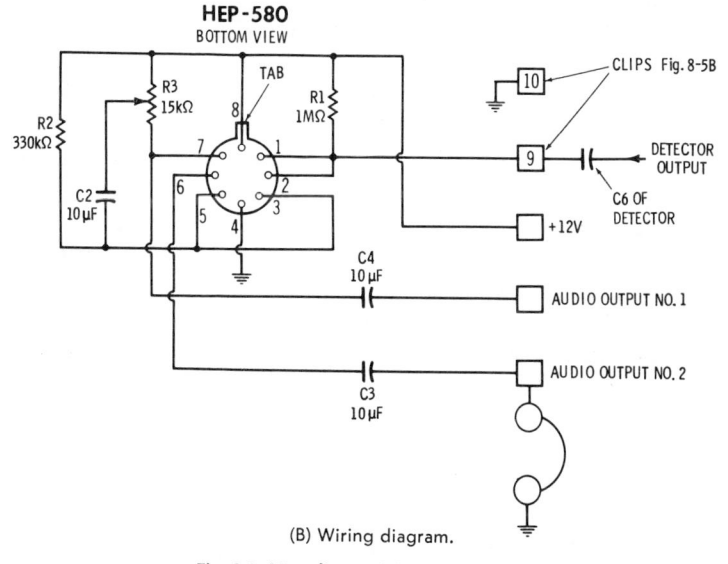

(B) Wiring diagram.

Fig. 8-8. IC audio-amplifier demonstration.

2. No definite sequence of wiring is necessary, although it is good practice to do so in an orderly manner to avoid mistakes. The ground clip of the regenerative detector output can be used as a common for the outputs of the integrated

circuit. Three additional clips are needed for the IC supply voltage (positive), and outputs No. 1 and No. 2.
3. Connect resistor R1 between the supply-voltage clip and soldering lugs 1 and 2. Connect another lead between these lugs and output clip 9 of the regenerative detector.
4. Connect a lead between lug 8 and the supply-voltage clip. Connect the outer terminals of the potentiometer between the supply-voltage clip and lug 7. Connect capacitor C4 between lug 7 and output No. 1 clip.
5. Connect capacitor C2 between the arm of the potentiometer and lugs 3 and 5. Connect resistor R2 from lugs 3 and 5 to the supply-voltage clip.
6. Connect lug 4 to the output ground clip. Connect capacitor C3 between lug 6 and output No. 2 clip. This completes the wiring of the headset amplifier. The headset itself is connected between output clip No. 2 and common.
7. Insert coil 1, which covers the high-frequency end of the broadcast band, into the coil socket of the detector. Wear the headset and tune in a broadcast station. If the output in the headset is very strong and distorted, you will have to back off the setting of the IC volume control, R3. Try out the other bands.
8. Connect audio output No. 1 clip to the input of the speaker audio amplifier. Strong volume level can now be obtained. In receiving strong signals, it will be necessary to lower the setting of the volume control that is part of the input circuit of the audio amplifier (amplifier constructed for Chapter 3 demonstration). This method of operation permits strong speaker output even in the reception of weak cw signals.

Index

A

Ac and dc electricity, 7-8
Ac beta, 46-47, 67
Ac electricity, 8-9
  average value, 9
  effective value, 9
  peak-to-peak value, 9
  peak value, 9
  root-mean-square (rms), 9
Ac operation, FET, 52
Ac-to-dc power supply, 56-62
Amateur and short-wave radio reception, 135-136
Ampere, 8
Amplifier, audio, 72-73
  at work, 80-81
  integrated-circuit, 136-140
  push-pull, 74-76
  two-stage, 76-80
Amplitude modulation (a-m), 83-84, 116-119
Antenna requirements, 87-88
Armstrong oscillator, 110-111
Audio amplifier, 72-73
  at work, 80-81
  integrated-circuit, 136-140
  push-pull, 74-76
  two-stage, 76-80
Audio oscillator, 93-103
  feedback, 93-96
  frequency of operation, 96
  types, 96-98
Average value, ac electricity, 9

B

Base, 44
  modulation, 117-119
Bias
  forward, 43-46
  reverse, 43-46
Bipolar transistor, 44-46
  ac beta, 67
  base, 44
  collector, 44
  dc beta, 66
  emitter, 44
  operation, 62-68
  voltage gain, 67-68

C

Capacitive reactance, 33
Capacitors and resonant circuits, 31-38
Carbon microphone, 70-71
Carrier, radio-frequency, 83
Cartridge
  ceramic, 71-72
  crystal, 71-72
Channel, 51
Code
  Morse, 100-103
  practice, 100-103
  reception with regenerative detector, 126-128
Collector, 44
  modulation, 119-120
Colpitts oscillator, 105-106
Coupling, inductive, 26
Crystal-controlled oscillator, 107-108
Crystal microphone, 70-71
Crystal radio, 88-92
Current
  electric, definition of, 7
  gain
    ac, 46-47, 67
    dc, 46, 66
    small-signal, 46-47, 67

D

d'Arsonval meter movement, 11
Dc and ac electricity, 7-8
Dc and ac operation, 46-48
Dc beta, 46, 66
Dc, pulsating, 40
Depletion area, 50
Detection, 86-87
Detector
  FET regenerative, 129-134
  regenerative, 125-128
  transistor, 123-125
Diaphragm, microphone, 70
Dielectric, 32
  constant, 33
Diodes, 85-86
Doping, 50

Drain, 51
Dynamic microphone, 69-70

E

Effective value, ac electricity, 9
Electric
  ac, 8-9
  ac and dc, 7-8
  current, definition of, 7
  lines of force, 31-32
  power, 8
  prefixes and multipliers, 8
  units and symbols, 8
Electromagnetic induction, 24-26
Electron, 7
Electrostatic lines of force, 31-32
Emitter, 44
  modulation, 119-120
Envelope, modulation, 117-120

F

Farad, 32
Feedback, 93-96
  in-phase, 95
  out-of-phase, 95-96
  polarity of, 94-96
    negative, 95-96
    positive, 95
  radio-frequency, 105-106
Field-effect transistor (FET), 48-52
  ac operation, 52
  channel, 51
  drain, 51
  gate, 51
  regenerative detector, 129-134
  rf oscillator, 110-113
  source, 51
Field strength, magnetic, 23-24
Filtering, 42-43
  inductance-capacitance, 42
  resistance-capacitance, 42
Flux, magnetic, 10, 23-24
Forward bias, 43-46
Frequency, 9
  measurement and calibration, 113-115
  resonant, 34
Full-wave rectifier, 41-42

G

Gate, 51

H

Half-wave rectifier, 41-42
Hartley oscillator, 105-106
Heat sink, 63
Holes, 43

I

Inductance, 25
Inductance-capacitance filter, 42
Induction
  electromagnetic, 24-26
  mutual, 26
  self, 26
Inductive coupling, 26
Inductive reactance, 25-26
Inductors and transformers, 23-31
In-phase feedback, 95
Integrated-circuit audio amplifier, 136-140

J

Junction
  pn, 44
  semiconductor, 43-44

L

Lines of force
  electric, 31-32
  magnetic, 10, 23-26

M

Magnetic
  field, 23-26
  flux, 10, 23-24
  lines of force, 10, 23-26
Meter
  movement, 10-12
    d'Arsonval, 11
    permanent-magnet, moving-coil, 11
    Weston, 11, 12
  sensitivity, 11
  VOM, 9-12
Microphone, 69-71
  carbon, 70-71
  crystal, 70-71
  diaphragm, 70
  dynamic, 69-70
  piezoelectric principle, 70-71
Modulation
  amplitude, 116-119
  envelope, 117-120
  methods, transistor, 119-120
    base modulation, 117-119
    collector modulation, 119-120
    emitter modulation, 119-120
  transformer, 119
Morse code, 100-103
Moving-coil cartridge, 71-72
Multipliers, electrical, 8
Mutual induction, 26

N

Negative feedback, 95-96
Negative reactance, 33
Npn transistor, 44-46
N-type semiconductor, 43

O

Ohm's law, 7-8
Oscillator
  audio, 93-103
  feedback, 93-96
  frequency of operation, 96
  types, 95-96
  radio-frequency, 104-115
    Armstrong, 110-111
    Colpitts, 105-106
    crystal-controlled, 107-108
    Hartley, 105-106
    tuned-base, 106-107
    tuned-collector, 106-107
    zero-beating, 108-110
  tone, 98-100
Out-of-phase feedback, 95-96

P

Peak-to-peak value, ac electricity, 9
Peak value, ac electricity, 9
Permanent-magnet, moving-coil
  meter movement, 11
Permanent-magnet speaker, 73-74
Permeability, 24
Phono cartridge, 71-72
  ceramic, 71-72
  crystal, 71-72
  moving-coil, 71-72
Piezoelectric principle, 70-71
Pn junction, 44
Pnp transistor, 44-46
P-type semiconductor, 43
Positive feedback, 95
Positive reactance, 33
Power, electrical, 8
Power supply, ac-to-dc, 56-62
Prefixes, electrical, 8
Primary winding, transformer, 26-27
Projects
  ac-to-dc power supply, 56-62
  amateur and short-wave radio
    reception, 135-136
  audio amplifier at work, 80-81
  bipolar transistor operation, 62-89
  capacitors and resonant circuits, 31-38
  code practice, 100-103
  crystal radio, 88-92

Projects—Cont'd
  FET regenerative detector, 129-134
  FET rf oscillator, 110-113
  frequency measurement and
    calibration, 113-115
  inductors and transformers, 23-31
  integrated-circuit audio amplifier, 136-140
  rectifier operation, 52-56
  tone modulation, 120-121
  tone oscillator, 98-100
  two-stage audio amplifier, 76-80
  using the VOM, 15-22
  wireless player, 122
Pulsating dc, 40
Push-pull amplifier, 74-76

R

Radio
  amateur and short-wave reception, 135-136
  carrier, 83
  crystal, 88-92
  frequencies, 82
Radio-frequency oscillator, 104-115
  feedback, 105-106
  types, 106-108
    Armstrong, 110-111
    Colpitts, 105-106
    crystal-controlled, 107-108
    Hartley, 105-106
    tuned-base, 106-107
    tuned-collector, 106-107
    zero-beating, 108-110
Reactance
  capacitive, 33
  inductive, 25-26
  negative, 33
  positive, 33
Receiver principles, 84-88
  antenna requirements, 87-88
  detection, 86-87
  diodes, 85-86
  tuning, 84-85
Record player, wireless, 122
Rectifier
  full-wave, 41-42
  half-wave, 41-42
  operation, 39-42, 52-56
Regeneration, 123
Regenerative detector, 125-128
  code reception with, 126-128
  FET, 129-134
Resistance, 7
Resistance-capacitance filter, 42
Resonant circuits and capacitors, 31-38
Resonant frequency, 34
Reverse bias, 43-46

143

R

Ripple, 43
Root-mean-square (rms), 9

S

Secondary winding, transformer, 26-27
Self induction, 26
Semiconductor
holes, 43
junction, 43-44
n-type, 43
p-type, 43
Sensitivity, meter movement, 11
Sine wave, 8-9
Small-signal beta, 46-47
Source, 51
Speaker operation, 73-74
Step-down ratio, transformer, 27
Step-up ratio, transformer, 27
Symbols, electrical, 8

T

Thermal runaway, 66, 75
Tickler coil, 126
Tone modulation, 120-121
Tone oscillator, 98-100
Transformer, 26-27
modulation, 119
primary winding, 26-27
secondary winding, 26-27
turns ratio, 27
step-down, 27
step-up, 27
Transformers and inductors, 23-31
Transistor
bipolar, 44-46
ac beta, 67
base, 44
collector, 44
dc beta, 66
emitter, 44
operation, 62-68
voltage gain, 67-68
detectors, 123-125
field effect (FET), 48-52
forward bias, 43-46

Transistor—Cont'd
modulation methods, 119-120
base modulation, 117-119
collector modulation, 119-120
emitter modulation, 119-120
npn, 44-46
pnp, 44-46
reverse bias, 43-46
unipolar, 48-52
ac operation, 52
channel, 51
drain, 51
gate, 51
source, 51
Tuned-base oscillator, 106-107
Tuned-collector oscillator, 106-107
Tuning, 84-85
Turns ratio, transformer, 27
step-down, 27
step-up, 27
Two-stage audio amplifier, 76-80

U

Unipolar transistor, 48-52
ac operation, 52
channel, 51
drain, 51
gate, 51
source, 51
Units, electrical, 8
Using the VOM, 15-22

V

Volt, 8
Voltage gain, bipolar transistor, 67-68
VOM, 9-12

W

Watt, 8
Weston meter movement, 11, 12
Wireless player, 122

Z

Zero-beating, 108-110

144